SCIENTOCRACY

SCIENTOCRACY

THE TANGLED WEB OF
PUBLIC SCIENCE AND PUBLIC POLICY

EDITED BY   PATRICK J. MICHAELS   TERENCE KEALEY

Contributions from   Trevor Burrus   Edward J. Calabrese
Jason S. Johnston   Terence Kealey   Ned Mamula   Patrick J. Michaels
Michelle Minton   Jeffrey A. Singer   Thomas P. Stossel

ISBN: 978-1-948647-49-6
eISBN: 978-1-948647-50-2

Cover design: Jon Meyers.
Interior Design: Westchester Publishing Services.
Printed in Canada.

Library of Congress Cataloging-in-Publication Data available.

1000 Massachusetts Ave., NW
Washington, DC 20001

INTRODUCTION

Science is the time-honored key to objective knowledge. Some of that knowledge—for example, about nutrition, climate change, hydrology, geology, or ecology—affects what we do every day. Science also informs governments that seek to define risks and mitigate dangers. The popular notion is that science is a force for good, that knowledge, derived from theory and experiment, gives rise to technological advancement, which results in improved lives for all.

We believe that this is not always the case. Science *can* be a force for good, and it has enhanced our lives in countless ways, but even a cursory look at the 20th century shows that what passes for science can be detrimental. This book documents only some of the more recent abuses of science that informed members of the public should be aware of.

Democratic states can not only adopt scientific error but, with government support, they can also help to embed and proselytize it. The mechanism is simple: peer review. It is on the basis of peer review that research grants are awarded, papers are accepted for publication, and promotions are determined. Once a research elite has adopted a scientific paradigm, that elite can harness peer review to reinforce its paradigm by excluding dissenting voices from federal funding, publication, and promotion.

As detailed in Chapter 1, President Dwight Eisenhower clearly understood the potential for government funding to entangle science and policy, both through funding and through other methods. Government funding sustains failing paradigms, and that artificial sustenance—when coupled with the pressure to publish—degrades science. Today, scientific papers are being withdrawn in increasing numbers because flawed findings passed review, inconvenient data were ignored, and results were fabricated. Positive findings—those that confirm a researcher's hypothesis—are increasing, something that should be astronomically improbable, but which may be related to the number of withdrawn works. And it is noteworthy that the tested hypotheses found to be supported by the data are usually the same ones that served as the basis for the research proposal that paid for the work. Why? Because reporting results that do not support a funded hypothesis can threaten the recipient's continued funding.

We also see another tactic being used to leverage more grant money from Congress—namely, the exaggeration of threats. Since the 1960s, environmental scientists have sequentially forecast existential threats from overpopulation, pollution, extinctions, global cooling, acid rain, global warming, stratospheric ozone depletion, low-level ozone increases, and fine-scale particulates. Ocean acidification looms on the horizon as the next threat.

The federal government also wrongly speaks in the name of science to achieve some policy goals. This book extensively details two dietary controversies, over fat and over salt, in which politics clearly trumped science, as well as two resource extraction issues, the Pebble mine in Alaska (Chapter 9) and uranium mining in Virginia (Chapter 8), in which the government—in the form of the Environmental Protection Agency in Alaska, and the National Research Council in Virginia—clearly cherry-picked science to fulfill political ends.

In Chapter 1, we examine the hypothesis that government funding of science stimulates economic growth. There's not much evidence for this. We are also led to believe that government funding of basic biomedical science promotes health in the United States. That has not happened either: an already long-running decline in death rates did not accelerate after the expansion of the National Institutes of Health in 1948. Advances in medicine, it seems, cannot be forced by funneling research money to pure scientists isolated from the bedside or the market; rather, medical advances emerge from the collaboration of clinicians, foundations, and pharmaceutical companies in the joint exploitation of adventitious observations (Chapter 6).

Nutrition

Legislative meddling, as well as government research funding, precipitated wars on fat and salt; these wars were costly, and they harmed American health (Chapters 2 and 3). The great U.S. postwar health scare was the epidemic of heart attacks, which came seemingly out of nowhere; at its height in 1968, it accounted for over 37 percent of all deaths in the United States. By contrast, all cancers then accounted for only 17 percent of all deaths, and strokes a mere 10 percent. Accidents, at 6 percent, were the fourth most common cause of death. At the time, no one could explain the epidemic of heart attacks, though it was probably—*probably*, for we still do not know for sure—caused by a combination of cigarette smoking, metabolic syndrome (the co-occurrence of high blood pressure, belly fat, and elevated blood glucose), and perhaps even infection with *Helicobacter pylori*, now known to be the cause of stomach ulcers. Research on this bacterium's effects on cardiovascular health continues today.[1]

Since time immemorial, mysterious epidemics have been falsely "explained" by brahmins with special knowledge, and the coronary epidemic was no exception. Two extremely influential (and therefore well-funded) scientists—Ancel Keys of the University of Minnesota and Lewis Dahl of the Brookhaven National Laboratory—asserted, respectively, that heart disease was caused by fat and salt in the diet.

We now know that these explanations represented unwarranted and even dangerous extrapolations from partial data. Moreover, equally credible and authoritative people at the time showed this to be so. But the prospect of a balanced debate within the scientific community was destroyed in January 1977 when the Senate Select Committee on Nutrition and Human Needs published its *Dietary Goals for the United States*, which stated that the science was settled: Americans needed to eat less saturated fat, less salt, and more carbohydrates.[2]

When challenged on the strength of the evidence, the chairman of the committee, Sen. George McGovern (D-SD), said, "Senators do not have the luxury that the research scientist does of waiting until every last shred of evidence is in," which is the opposite of the truth: research scientists are at leisure—and are perhaps even obligated—to explore every possible hypothesis, but senators should not issue advice *until* every last shred of evidence is in, because they may otherwise issue misleading or even dangerous advice—as they did, beginning in 1977.

Federal agencies adhere to government policy, and the executive branch (in the shape, jointly, of the U.S. Department of Health and Human Services and the U.S. Department of Agriculture [USDA]) took on the responsibility of publishing the Senate committee's wrong health advice in "Dietary Guidelines for Americans," which was updated every five years and popularized by the USDA's Food Guide Pyramid of 1992, MyPyramid of 2005, and MyPlate of 2011. These reinforced the anti-fat, anti-salt, pro-carbohydrate message.

In 1981, Rep. Albert Gore (D-TN), as chairman of the House Committee on Science Subcommittee on Oversight, became interested in salt. He opened two days of hearings on the topic by saying, "We can begin to see a consensus emerging in the research community about the reduction of salt intake for the average American," language eerily similar to the rhetoric he would later apply to global warming.[3] Joining him in his anti-salt certainty was Rep. Henry Waxman (D-CA). Gore and Waxman ultimately became two of the most prominent political proponents of climate science alarmism as well.

The convergence of politicians and powerful scientists to create the image of "doing something" about a "problem" is repeated throughout this book. It has consequences both for regulatory policy, which affects ordinary Americans, and for science policy, which determines the shape of future funding and research.

Under the influence of a scientific elite, the very same politicians who declared the science in nutrition to be settled would proceed to make the same claim about climate change.

That the world is warming is unquestioned, and that anthropogenic carbon dioxide (CO_2) is a cause is also unquestioned, but the damage that anthropogenic CO_2 will cause cannot at present be quantified with confidence. Depending on the rate, quantity, and timing of future warming, and accounting for the opportunity costs of action, cost estimates for carbon dioxide emissions can range all the way from expensive to negative (i.e., some have argued that CO_2 emissions have benefits).[4]

In their climate alarmism, many politicians have of course followed the lead of the scientific community that they funded. Literature summaries produced in the United States by the Global Change Research Program and the Environmental Protection Agency (EPA), and internationally by the United Nations' Intergovernmental Panel on Climate Change, reflect the fruits of that funding. These summaries include predictions based on computer models known as general circulation models (GCMs) (Chapter 10).

In making predictions, the climate science community uses GCMs, in preference to exploiting history. Is all of Greenland's ice about to melt, causing sea level to rise by 20 feet? Reality argues not. Historical studies, not models, show that roughly 118,000 years ago, and continuing for 6,000 years, increased solar radiation caused high-latitude summer temperatures over the icecap to be 6–8°C (10–14°F) warmer than the 20th-century average, yet Greenland lost only about 30 percent of its ice.[5] This is greater integrated heating than humans could exert there, and yet the fear of a Greenland-induced disaster lives on.

In fact, GCMs have a terrible track record. Concerning the three-dimensional warming of the climatically critical tropical latitudes, only one of 32 modeling groups correctly predicted the climatic behavior of the past 35 years. Notably, that model, the Russian INM-CM4, predicts the least global warming of them all. A rational policy community would use the Russian model as the operational forecast model, but instead the EPA and other agencies use the average and the dispersion of all 32 groups when assessing future climate change impacts.

Systematic errors are probably occurring because the models are erroneously "tuned" to simulate the warming from 1910 to 1945. The carbon dioxide concentration in 1910 was far too low to induce the magnitude of warming that followed. Indeed, Frederic Hourdin, the head of the French national modeling team, recently called for systematic documentation and rationales for the "tuning" of the models.[6]

Pollution and Ionizing Radiation

Chapter 11 details the slipshod science used by the EPA to justify its regulation of fine-scale particulates, which are ubiquitous in the atmosphere. One contributing factor is that the scientists who have received EPA grants are the same people who advise it on policy.

Not everybody loved EPA administrator Scott Pruitt, but on October 31, 2017, soon after his appointment, he revealed that scientists on just three of the EPA's science advisory boards had, over the previous three years, collectively received research grants from the agency totaling $77 million.[7] The scientists were, in short, hardly independent of the EPA; rather, they had a compelling interest in its ongoing campaign of pollution alarmism.

What about the effects of ionizing radiation and carcinogens? We know that the more radiation we are exposed to, the greater the hazards, so it makes sense to regulate to achieve radiation-free lives, right?

Here is a secret that is rarely disinterred from the deepest recesses of the more obscure leaves of the scientific literature: a little radiation, and possibly a small exposure to fine particulates, can be good for you. In a process called hormesis, mild exposure to dangerous agents stimulates repair and antioxidant responses that leave a person stronger than before. In more technical language, dose-response curves are not linear but biphasic (Chapter 7).

Scientists at the regulatory agencies have not yet proclaimed this good news. But in 2018, President Donald Trump's EPA began to consider the hormesis model in formulating regulations. Previously, government capture of an important branch of science—regulatory science—was based on a failed model that assumed there is no safe threshold—for example, for radiation exposure.

Drugs

The federal government's casuistry over drugs is revealed by the approach the Drug Enforcement Agency (DEA) takes to research—namely to suppress it. As new recreational drugs emerge, the agency assigns them (sometimes on the basis of very limited knowledge) to Schedules I through V, with Schedule I substances purportedly being highly addictive and with no redeeming value. This categorization renders further federally supported research on Schedule I drugs almost impossible; private research is likewise difficult or impossible because the substances are illegal. Without having to prove its assertions, the agency can declare drugs to be more harmful than the devastation their prohibition wreaks on minority and other vulnerable sectors in the United States.

Which matters. Prohibition between 1920 and 1933 harmed communities of all ethnicities, and those communities needed no research to tell them that Prohibition's impact was more destructive than alcohol's; hence the passage of the 21st Amendment. But the War on Drugs disproportionately destroys minority communities. The long-run decline of poverty among African Americans stopped in 1969,[8] the year the War on Drugs was declared at the federal level (though some states had opened hostilities sooner); and if the majority is not informed, by research, of the disproportionate damage its legislation wreaks on minorities, then—in all good faith—it will continue to legislate for prohibition.

A cost-benefit analysis can be made only on the basis of firm data, which renders the DEA's reluctance to facilitate research dismaying. But truth is the first casualty of war, a maxim that applies in this case to the actual prohibition of research on Schedule I drugs (Chapter 5).

Opioids

The damage caused by opioids is largely caused by their prohibition, but since that damage is focused on the victims themselves, the wider community—who are kept in ignorance of the minuscule number of people who graduate from legitimate prescription use to chronic addiction—will continue to press for their prohibition (Chapter 4).

Conclusion

Scientific research is the time-honored key to objective knowledge. In the past it was funded pluralistically, but today certain portions of the market for knowledge are dominated by a single buyer—the government. This is especially true in the research fields that impinge on the regulatory sphere, such as pollution and climate change. As discussed in Chapter 1, science today is in systematic trouble.

Everyone knows about feedback loops, often known informally as "vicious cycles," but in science we now see the *feed-forward loop* of bad science creating paradigms and policies that refuse to die. Such loops will be unwound only by attenuating the bond between science and the state. We may not know what research will find when the feed-forward loop is broken, but that's exactly why it should be broken. Research *should* surprise us.

Please don't let us make you into cynical coots, however. Science has done much for us under public funding. Many fundamental questions have indeed been answered, including, particularly in physics, questions that otherwise might not yet have been answered. We look forward to a future of still more vigorous scientific discovery; we ask only that it be structured in a more voluntary and polycentric manner, and we believe that the chapters you are about to read will more than justify this desire.

SCIENCE AND LIBERTY
A Complicated Relationship

S cience has traditionally been represented as an agent of liberty, and among the tyrants from which it liberated us were the Church and poverty.

The Church

Science first emerged in classical Greece, and though the Greek city-states were hardly theocracies, nonetheless tensions emerged between the priests, who drew their power from received authority, and the scientists, who drew their insights from reason and observation. And when there were victims, they were not the priests. Thus when, during the 5th century BCE, Anaxagoras claimed the sun might not be a god but rather "a rock bigger even than the whole Peloponnese," the Athenians would have executed him had his old friend Pericles not engineered his escape to Lampascus.

It was therefore impressive of Hippocrates, known as the father of medicine, to write during the 4th century BCE in *On the Sacred Disease*:

I am about to describe the disease called "sacred" [epilepsy]. It is not in my opinion any more divine or sacred than other diseases but has a natural cause, and its supposed divine nature is due to men's inexperience and to their wonder at its peculiar character.

The priests in classical Greece were not, however, as powerful as those who emerged in medieval Europe, which is why the writings of Francis Bacon still have the power to startle. Bacon (1561–1626), an English lawyer and politician, is recognized as the first great philosopher of science, and though he lived in an age of faith, he nonetheless dared to portray science, not religion, as a portal to the sublime, writing in 1607 in *Cogitata et Visa de Interpretatione Naturae* ("Thoughts and Conclusions on the Interpretation of Nature"),

> It is this glory of discovery that is the true ornament of mankind[1] . . . the improvement of man's mind and the improvement of his lot are one and the same thing.[2]

It was also Bacon who wrote *ipsa scientia potestas est*, "knowledge itself is power," which—in an age when power was the perquisite only of Church and state—was a radical statement. Certainly the Church (or perhaps we should write churches, because the Reformation had inadvertently helped create intellectual space for science) suppressed challenges to its intellectual authority. This is why we still remember that in 1633, Galileo was shown the instruments of torture by Pope Urban VIII, and that he was thereafter sentenced to house imprisonment for believing the earth rotated round the sun.

Although the Reformation had empowered dissent from Catholic doctrine, the reformers could be equally intolerant, and their intolerance extended to science. Thus Martin Luther condemned as a "fool" the astronomer who believed the earth rotated round the sun, saying:

> So it goes now. Whoever wants to be clever must agree with nothing that others esteem. He must do something of his own. This is what that fellow does who wishes to turn the whole of astronomy upside down. Even in these things that are thrown into disorder I believe the Holy Scriptures, for Joshua commanded the sun to· stand still and not the earth.[3]

Calvin also denounced believers in heliocentricity as "stark raving mad" and "possessed by the Devil."[4] And he was a man not to be crossed: he did, after all, burn Servetus alive, and Servetus was a great scientist who had discovered pulmonary circulation (though to be accurate, Calvin burned him for religious rather than scientific reasons, because Servetus had questioned the doctrine of the trinity).

Nonetheless, for all its limitations, the Reformation broke the Roman Catholic monopoly on truth, and as medieval thinking yielded to the Age of Reason (or Enlightenment) of the 18th century, so science advanced at the expense of religion. Today, science's victory over religious intolerance is nearly complete in the West, and although obscurantism still rules in some parts of the world, the advance of scientific reason seems inexorable. Yet even as science has flourished in the modern era, it has lately come to be captured to a great degree by the state.

One illiberal aspect of present-day science is that scientists have long sought state funding and have therefore long aligned themselves with state doctrines. The perennial anxiety of scientists for funding—from any source—was parodied as long ago as 1726 by Jonathan Swift in his satirical take on the fellows of the Royal Society of London:

The first man I saw was of a meagre aspect, with sooty hands and face, his hair and beard long, ragged, and singed in several places. His clothes, shirt, and skin, were all of the same colour. He had been eight years upon a project for extracting sunbeams out of cucumbers, which were to be put in vials hermetically sealed, and let out to warm the air in raw inclement summers. He told me he did not doubt that in eight years more he should be able to supply the Governor's garden with sunshine at a reasonable rate; but he complained that his stock was low, and he entreated me to give him something as an encouragement to ingenuity, especially since this had been a very dear season for cucumbers. I made him a small present, for my lord had supplied me with money on purpose, because he knew their practice of begging from all who go to see them.[5]

The sentiment had been expressed less humorously a century earlier, in 1605, by Francis Bacon himself when he told King James I that "there is not any part of good government more worthy than the further endowment of the world with sound and fruitful knowledge."[6]

Bacon was therefore not only a great philosopher of science; he was also the first to argue that science was a public good that required public funding. Science, Bacon wrote, is a "universality" that benefits everyone, not any particular individual: "The benefits inventors confer extend to the whole human race."[7] That is the classic description of a public good. The individual who makes a pen to sell to another individual will receive from the sale of that private good the full profit, but the individual who invents the *idea* of the pen will not receive the full profit from all the subsequent sales of pens across the globe and across the ages. Therefore, Bacon said, no one will invent a new technology, for which, he said, "there is no ready money."[8]

It is a false argument because it ignores the principle of *opportunity benefit*, which is the converse of opportunity cost. If there is a choice between doing A or B, and if A is chosen over B, the opportunity cost is the forgone benefit from B. Yet if A is more valuable than B, it is rational to choose A for its additional or opportunity benefit. Well, if the individual who invents the pen has, as his forgone opportunity, the benefit of ploughing a field, say, and if the benefit from ploughing the field is $10 in profit, and if the benefit of inventing the pen is $20 in profit, that individual will rationally invent the pen even if the total global profit from the sales of pens is $10 billion. The inventor of the pen may make only $20 in profit from the pens he sells, and this may be only a fraction of the global profits of the whole pen market of $10 billion. But because it's $10 more than he earns from ploughing a field, the individual will still be motivated to invent the pen.

The man who contested Bacon's public-goods argument was a Scot. Adam Smith (1723–1790) was eager, 150 years after Bacon had proposed that science was a public good, to test Bacon's idea against experience. Smith recognized public goods as being "of such a nature that the profit could never repay the expense to any individual or small number of individuals," but he found that industrial technology did not fit that category and that the profit did repay the expense to individuals.[9] So in 1772–1773, in his *Lectures on Jurisprudence*, he wrote,

If we go into the workplace of any manufacturer and . . . enquire concerning the machines, they will tell you that such or such a one was invented by common workman.[10]

In Smith's footsteps, Marx and Engels also found science to be endogenous to markets, writing in the *Communist Party Manifesto* of 1848 that,

The bourgeoisie, during its rule of scarce one hundred years, has created more massive and more colossal productive forces than have all the preceding generations together.[11]

And in 1942, Joseph Schumpeter wrote,

Industrial mutation incessantly revolutionizes the economic structure *from within*. [Schumpeter's emphasis][12]

The evidence of industrialization, therefore, suggested to empirical observers that governments need not fund science.

The Linear Model

Bacon also suggested that industrial science emerged from academic science, and in *The Advancement of Learning* he wrote,

If any man think philosophy and universality [science] to be idle studies, he doth not consider that all professions [technology] are from thence served and supplied.[13]

It was Bacon, therefore, who proposed the so-called linear model, which looks like this:

But Smith dissented, believing that in practice it was advances in industrial technology that stimulated academic research, "The improvements which in modern times have been made in several different parts of philosophy [science] have not, the greater part of them, been made in universities."[14] So Adam Smith's model was:

Smith did not, of course, rule out the possibility that science could feed into new technology, writing that some "improvements in machinery have been made by those who are called philosophers or men of speculation." But he reported it was a much less important source of new technology than preexisting technology.[15] And modern scholarship confirms Smith's reports: even modern science is only marginally important to industry. In 1991, Edwin Mansfield of the University of Pennsylvania surveyed 76 major U.S. firms that collectively accounted for one-third of all sales across seven key manufacturing industries, finding that only 10 percent of their innovations emerged from recent academic research—and those innovations were marginal ones, accounting for only 3 percent of sales and 1 percent of the savings or profits to industry resulting from innovation.[16]

Escape from Poverty: Testing the Models

Thus, the empirical evidence suggests today, as it suggested in the past, that governments need not fund science. Yet as Smith's friend David Hume once noted, "All human affairs are entirely governed by opinion,"[17] and during the 18th and 19th centuries the West engaged in a natural experiment: Britain, in its science policies, subscribed to Adam Smith's model of scientific laissez faire, while the autocratic governments of France and the German states subscribed to Francis Bacon's model of scientific dirigisme. And the canonical Industrial Revolution occurred in a country, Britain, whose governments did not generally fund research designed to support the market.

The British government funded mission research, which was research in aid of its geopolitical ambitions. The classic example was the search for an accurate method of determining longitude following the 1707 Scilly Isles Royal Navy disaster. But the British government also did not generally fund research designed to correct perceived market failure.

The French government did, however, and it funded, among other projects, the Jardin du Roi (1626; for medicinal plants, botany, and pharmacy); the Académie des sciences (1666; the Royal Society in London, by contrast, founded in 1660–1662, received no state support); its *Journal des sçavans*; the Ecole de Rome (1666; painting, architecture, and design); the Académie royale d'architecture (1671); the Ecole des ponts et chaussées (1716; civil engineering); the Ecole royale du génie (1748; military engineering); the Ecole gratuite de dessin (1767; drawing and design); the Ecole des mines (1778); the Ecole polytechnique (1794); and three chemistry research laboratories (in the King's library, the Louvre, and the Observatoire). The German states, too, supported their universities and *technische Hochschulen*. Yet the Industrial Revolution was British, not French or German.

Few natural experiments in economics are clean. The French government also eliminated incentives for manufacturing from its tax system. And the British government did fund some basic science. For example, for 10 years between 1803 and 1813, it funded an annual course of lectures in agricultural chemistry delivered by Humphry Davy, though they were discontinued for lack of interest; during the 1820s, it also funded Charles Babbage's computer experiments, though those failed; and the Scottish government did subsidize its universities, though not comprehensively. Notwithstanding, the contrast between the philosophies of the government in the United Kingdom and the governments of France and the German states is clear: the Continent followed Bacon in supposing markets failed in science, the UK followed Smith in supposing they did not.

The United States was another country whose government, until recently, primarily funded only mission research. It did fund science, temporarily, during wartime. Thus, the Civil War saw the founding of the National Academy of Sciences (NAS) to help design ironclads and other military materiel;[18] World War I saw the creation of the National Research Council to advise on the synthesis of poison gases and other weapons; and World War II saw the creation of the Office of Scientific Research and Development (OSRD), which spawned, among other things, the atom bomb. But these were funded only during the actual fighting, and their funding was stopped with the return of peace.

There were only two exceptions to the story of American prewar research laissez faire. One was mission research, which is the kind of research necessary to further specific government actions. Over the decades, the federal government established an increasing number of institutions to further mission research, including the Library of Congress in 1800 and the Coast Survey in 1807, and the execution of their responsibilities required research. The National Institutes of Health (NIH), for example, can trace its origins back to the Marine Hospital of 1798. But these institutions of mission research were not Baconian in intention. They were not funded to correct some supposed systematic "market failure"; they were founded only to support specific government missions.

The second exception was the creation of the land-grant colleges, in 1862, under Abraham Lincoln. Farmers in America had long been poor because growing food was too easy, so they grew too much, driving down prices. As the economic historian Eric Jones wrote:

> European farming methods were preternaturally productive in the New World. Time and again European travellers complained that American farmers wasted manure. Their dung heaps rose to tower over the red Palatine barns of the colonies; why were they not spread on the land?[19]

They were not spread on the land because of the preternatural productivity of the New World. Likewise the land-grant colleges were not only economically irrational; they also crowded out the preexisting colleges. In vetoing an earlier version of the bill in 1859, James Buchanan wrote:

> This bill will injuriously interfere with existing colleges ... [which] have grown up under the fostering care of the states and the munificence of individuals ... what the effect will be on these institutions of creating an indefinite number of rival colleges sustained by the endowment of the federal government is not difficult to determine.[20]

Terry Reynolds, moreover, has condemned

> the traditional view [that] only with the passage of the Morrill Act in 1862 and the creation of the land grant college did engineering

find a firm place in academia [because] literally dozens of ante-bellum colleges in all parts of the nation taught practical sub-jects including engineering.[21]

Nonetheless, the political power of the farmers ensured the fed-eral government's support of their education and research. Yet as late as 1940 the federal and state governments between them funded only 23 percent of U.S. research and development (R&D)—and almost none of its basic science. Federal funding existed primarily for the two mis-sions of defense and agriculture.[22] By about 1890, however, the United States had already overtaken Britain in GDP per capita; so the globe's two leading countries in succession, the 19th-century United Kingdom and the 20th-century United States, achieved their leads largely under research laissez faire.

But while the United States under research laissez faire was the coun-try that overtook the United Kingdom, the French and German states failed during the 19th century even to converge on the UK's GDP per capita. By 1913, U.S. GDP per capita was 14 percent greater than the United Kingdom's, while French and German GDP per capita were both 12 percent less,[23] which suggests that the government funding of research does not produce an economic benefit.

It is a myth that Germany overtook the United Kingdom economically during the 19th century. That myth's extraordinary resilience is unfortu-nately too easy to understand. Yet its discrediting has not been a secret, as Table 1.1 and Paul Bairoch's GNP per capita data show.

TABLE 1.1
GNP per capita (1960 U.S. dollars)

	1830	1913
United Kingdom	370	1,070
West Germany	240	775
France	275	670
United States	240	1,350

Note: Paul Bairoch extrapolated backwards from the geographical expression that was West Germany in 1960.

Source: David S. Landes, *The Wealth and Poverty of Nations* (New York: Norton, 1998), p. 232.

Bairoch's data on the different national levels of industrialization are similar. Yet activists have long suggested that the funding of science by German governments allowed Germany to overtake Britain during the 19th century. This is a myth that powered the campaign for the government funding of science.

Moreover, following the postwar inauguration of considerable U.S. federal government funding for research, both pure and applied, there was no increase in the long-run U.S. rate of per capita GDP growth, which has been remarkably steady since 1830.[24] In a parallel development, the UK government's 1913 inauguration of systematic funding of research (its first research council, the Medical Research Council, was founded in 1913) witnessed no increase in the long-run rate of per capita GDP growth either; it too has been remarkably steady since 1830.[25]

When government-funded research is tested econometrically against contemporary national rates of per capita GDP growth, it fails to demonstrate economic benefit. In 2003, for example, an Organisation for Economic Co-operation and Development (OECD) study of the growth rates of the 21 leading world economies between 1971 and 1998 found "a significant effect of R&D [research and development] activity on the growth process," but it was only "business-performed R&D that . . . drives the positive association between total R&D intensity and output growth."[26]

The OECD was discomfited by its own findings ("the negative results for public R&D are surprising"), which, however, echoed earlier research by Walter Park of American University. Using published OECD data, Park found that the direct effect of the public funding of R&D on economic growth "is weakly negative, as might be the case if public research spending has crowding-out effects which adversely affect private output growth."[27]

Likewise, in 2007 Leo Sveikauskas of the U.S. Bureau of Labor Statistics, reviewed the accumulated, published, empirical econometrics data and concluded that "the overall rate of return to R&D is very large. . . . However, these returns apply *only* to privately financed R&D in industry" (Sveikauskas's emphasis).[28]

These findings by the OECD, Park, and Sveikauskas speak to another tyrant that research has slain: poverty and its associated distempers of disease and premature death. For millennia, humans lived a Malthusian existence, where any economic advantage achieved by a technical innovation was immediately swallowed by population growth that returned individuals to a condition of existential peril, stalked by starvation, disease, and premature death. The R&D of the Industrial Revolution, however, changed that, creating wealth at such a rate as to exceed and then shrink the birthrate, thus leading to our current historically unimaginable levels of wealth. But it was only privately funded R&D that achieved that near miracle. And government-funded R&D, by crowding out the private element, has endangered continued growth—and may endanger previous gains as well.

A recent natural experiment, moreover, confirms that government R&D crowds out private forms. Over the past 30 years the major countries of the OECD, in response to common budgetary pressures, have cut their funding of R&D (from an average of 0.82 percent of GDP to 0.67 percent), yet the concomitant rise in private R&D funding (from an average of 0.96 percent of GDP to 1.44 percent) has more than compensated for the cuts.[29] Crowding out is not the behavior of a public good.

Industrial Research Is Not Built on Free Access to Knowledge

The empirical evidence suggests, therefore, that governments need not fund science, and the reason is that knowledge is not freely available. Mansfield and his colleagues examined 48 products of major industries in New England that had been copied during the 1970s and reported that the costs of copying were on average 65 percent of the costs of innovation.[30] This was because the copiers had to rediscover for themselves the tacit knowledge embedded in the original innovation. That rediscovery can be so laborious that copying an innovation sometimes cost the copiers *more* than it had cost the original innovators. A survey by Richard Levin and colleagues of 650 R&D managers found similar results for the costs of industrial copying.[31]

Mansfield and colleagues and Levin and colleagues reported only the marginal costs of the actual copying. Nathan Rosenberg, as well as Wesley Cohen and Daniel Levinthal, have shown that, because of the tacitness of knowledge, companies seeking success in the market need first to sustain the fixed costs of a research staff whose activities are directed toward maintaining their own expertise in pure as well as applied science.[32] This is why companies cannot neglect even pure science: the Science Policy Research Unit at the University of Sussex in the UK showed that some 7 percent of industrial R&D is spent on pure science, and that some companies publish as much as medium-sized universities.[33] Consequently, as Paula Stephan wrote, "The research of some scientists and engineers in companies like IBM, AT&T, and Du Pont is virtually indistinguishable from that of their academic counterparts."[34] And since there is a positive correlation between the amount of pure science that companies publish and their profits, these costs will not be trivial.[35] Companies need to bear the costs of information.[36] And they also need to bear the costs of failed imitation attempts. We may therefore not know for certain what the average costs of copying in industry are, but they appear to be high enough to show that research, in industrial practice, is excludable in the sense that it is not available to free riders.

The fact that knowledge, including science, is tacit, was recognized from the outset of the modern scientific enterprise. Thus, in *De revolutionibus* (1543), Nicolaus Copernicus wrote that he was writing only for those qualified to read it: "Mathematics is written for mathematicians."[37] Technological tacitness, too, has been long recognized, and Diderot wrote in the *Encyclopédie* (1751–1766) that "in all techniques, there are specific circumstances relating to the material, instruments, and their manipulation that only experience teaches."[38] And Michael Polanyi noted in *Personal Knowledge* that the tacitness of knowledge "restricts the range of diffusion to personal contacts." It was also Polanyi who wrote that "we can know more than we can tell" and who described "the ultimately tacit nature of all our knowledge."[39]

The tacit nature of industrial knowledge has other important effects on the dissemination of innovations. On surveying 650 senior managers from manufacturing firms across more than 100 industries in

North America, Levin and colleagues reported that, on average, only three to five firms had the capacity to duplicate a typical innovation; presumably only these firms had the appropriate tacit knowledge at their disposal and were inclined to make use of it.[40] In science, too, most important advances are made by only a small proportion of scientists,[41] so the number of scientists who possess the relevant tacit knowledge to copy important discoveries in each field is also small. Craig Venter, for example, noted that in 1992, when venture capitalists (led by Rick Bourke of Dooney and Bourke, Inc.) and the public sector (led by James Watson of the Human Genome Project of the National Institutes of Health) were competing to fund the sequencing of the human genome, they had only two labs over which to compete—Venter's and John Sulston's.[42] These are not the characteristics of a public good.

Copying in academia can be just as expensive. For example, after the transversely excited atmospheric (TEA) pressure CO_2 laser emerged from Jacques Beaulieu's laboratory in 1970 as a significant advance in laser science, other laser research laboratories wanted to build one, but Harry Collins, a sociologist of science, found that no other laboratory had "succeeded in building a TEA laser using written sources."[43] Personal visits to preexisting TEA laboratories were essential.

But not anyone could visit. Collins found that existing laboratories would receive only visitors who had conducted prior research in the field, because it was deemed that others would not benefit. Moreover, "nearly every laboratory expressed a preference for giving information only to those who had something to give in return."[44] Visitors, therefore, had also to be contributors, engaged in a two-way exchange of information.

In showing that publication is not always enough to allow copying in science, Collins recapitulated the experience of Robert Boyle, the commanding spirit of the early Royal Society. Boyle, who is still remembered for his eponymous law (that there is an inverse relationship between the pressure and volume of a gas) focused his gas research on his air pumps, yet however carefully he described the pumps in his publications, the only people who could copy them were those who visited him personally and handled one themselves.[45]

The Contribution Good

We have shown that science cannot be a public good, yet it is not a private good either, since knowledge spreads—by publication, patents, or reverse engineering—in ways that material private goods do not. But knowledge spreads only between those who have access to it by virtue of their own contributions. Elsewhere, we have therefore modeled science as a *contribution good*, a good whose benefits are available only to those who have contributed to it.[46] Unlike other models, ours accounts for the takeoff of the Industrial Revolution.

The model also accounts for the spread of knowledge in ways that are mutually advantageous. One of the great myths is that today's companies are secretive, but the contrary is also true. Indeed, commercial success seems to depend on competitors *sharing* knowledge. An MIT survey of 11 steel companies in the United States found that 10 of them regularly swapped proprietary information,[47] and an MIT international survey of 102 firms found that no fewer than 23 percent of their important innovations came from swapping information with their rivals: "Managers approached apparently competing firms in other countries directly and were provided with surprisingly free access to their technology."[48]

Those MIT surveys were of mature industries, but among industries whose technology is still nascent, knowledge is not so much shared as jointly researched to mutual advantage.[49] Only as products get close to customers do companies turn truly competitive.

The Significance of Science Not Being a Public Good

At least two important lessons emerge from the foregoing. First, the scientific community has for decades promoted an economic model of science that is demonstrably false. As with much else in this book, it is an example of a scientific consensus that cannot be trusted.[50]

Second, the fact that governments need not fund science is enormously liberating, because it discredits an idea that had legitimized government intrusion into a major area of life—namely economic growth. Growth is of key importance, and it is based on innovation;[51]

and if innovation needs to be funded by government, then government can legitimately claim to be the source of economic growth. But if governments do not need to fund science to produce innovation, then lovers of freedom can legitimately withhold from government its involvement in that activity.

When Bacon wrote *ipsa scientia potestas est*, "knowledge itself is power," the entity he really wanted to empower was the state, not the individual researcher. In his essay "Of the true Greatnesse of Kingdomes and Estates," Bacon wrote that national greatness emerged from "the command of the Seas." And who then commanded the seas? The Iberians. In Bacon's day, the Portuguese and the Spanish dominated the globe. According to Bacon, the Portuguese had reached India, and the Spanish had reached America, thanks to science. In his legendary research institute in Sagres, Henry the Navigator, Bacon claimed, had researched the science that underpinned novel technologies, including the improved compasses, ships, cartography, and tools of navigation. With these the Portuguese had reached Africa and Asia, and the Spanish had reached America. "The West Indies had never been discovered if the use of the mariners' needle had not been discovered," Bacon wrote, while technology, he maintained, was the source of political power: "Printing, gunpowder and the magnet had altered the whole face and state of the world."[52]

But every "fact" that Bacon claimed about the Portuguese and Spanish, however, was false. The Portuguese and Spanish were not scientists: forged in the brutalities of the Reconquista, they were ruthless, Muslim-hating warriors who, scarred by the failures of the Crusades, sought to outflank their Islamic enemies by striking, via Africa, at their supply chains in Asia. As recent scholarship has confirmed, it was simply a myth—one well propagated by the early Portuguese chroniclers—that Henry the Navigator had founded a research institute in Sagres. The only thing he produced in Sagres was an illegitimate child,[53] and the technologies he employed in his so-called voyages of discovery were the conventional technologies of his day.

War has always justified the government funding of science, and World War II was no exception. Its mobilization of American science started in August 1939, when Albert Einstein signed a letter to President Franklin Roosevelt noting that, thanks to recent research, an "extremely powerful bomb of a new type may . . . be constructed," and warning that Germany had, ominously, prohibited all sales of the rare, recently discovered element uranium from the former Czechoslovakia, which only five months earlier it had invaded and occupied.[54] Thereafter the Manhattan Project, the U.S. effort to build an atomic bomb, grew into the world's largest scientific endeavor.[55]

Toward the end of the war, though, Vannevar Bush feared the apparently inevitable demobilization of the Office of Scientific Research and Development (OSRD), which he directed and which oversaw wartime science, so he solicited a famous letter from Roosevelt, written on November 17, 1944, though it could have been written by the ghost of Francis Bacon:

> There is, however, no reason why the lessons found in this experiment [OSRD and the Manhattan Project] cannot profitably be employed in times of peace. The research experience developed by the Office of Strategic Research and Development and by the thousands of scientists . . . should be used in the days of peace ahead for the improvement of the national health, the creation of new enterprises, and the betterment of the national standard of living.[56]

Roosevelt asked Bush four questions:

> How could we diffuse the scientific knowledge gained in the war effort?
> How could the advances in medicine related to the war, such as the widespread use of antibiotics, be continued in peacetime?
> What can Government do now and in the future to aid research activities . . . ? [the capitalization of the "G" may be telling]
> Can we develop and train new scientists so that the remarkable advances that took place in the war effort can be maintained?[57]

Roosevelt then closed with a Baconian flourish:

New frontiers of the mind are before us, and if they are pioneered with the same vision, boldness, and drive with which we have waged this war we can create a fuller and more fruitful employment and a fuller and more fruitful life.[58]

Meanwhile, in July 1945, Bush published *Science: The Endless Frontier*,[59] his blueprint for nationalizing basic science. Although his rationale was that a strong economy was impossible without government funding for pure or basic science, his main thrust was actually that a strong military was impossible without it. Bacon once again.

With the resumption of peace, though, wartime agencies such as the OSRD were, following the familiar playbook, demobilized; but in 1947 Congress—inspired in part by Bush's book—passed a bill to create a peacetime national science foundation. It was vetoed, though, by President Harry Truman because it "would, in effect, vest the determination of vital national policies, the expenditure of large public funds, and the administration of important governmental functions in a group of individuals who would be essentially private citizens."[60]

Congress's bill had envisaged a national science foundation that—rather like today's, actually—distributed government money to applicants whose work had been reviewed and deemed worthy by their fellow scientists. Truman, on the other hand, had envisaged an institution along the lines of the OSRD, which in effect was a branch of the federal government. But by 1950 the Cold War was heating up, thus creating a government demand for strategically relevant science.

In his farewell address, George Washington had warned against "permanent alliances" and against "excessive partiality for one foreign nation, and excessive dislike for another"; these statements are often summarized as his warning against "foreign entanglements." Until 1947—the World Wars excepted—avoiding foreign entanglements had mostly remained U.S. policy (the United States' refusal to join the League of Nations is an example). But in 1947 the British, nearly bankrupt, suspended Pax Britannica and prepared to abandon Greece and Turkey to Soviet incursions,[61] whereupon Truman codified his Doctrine (which was strengthened in 1948 and led to the creation of NATO in

1949), which might be summarized as Pax Americana. Isolationism was done, and America, like Britain before it, was now permanently at war with armed international transgressors: in 1949 the Soviet Union detonated its first fission bomb.

Modern war needs science, and modern war needs scientists. In 1942, 1943, and 1945, the U.S. Senate's Subcommittee on War Mobilization had held hearings on America's shortage of wartime scientists (the peacetime complement had been fully adequate for peace, but not for war), and it was that subcommittee's chairman, Harley M. Kilgore (D-WV) who subsequently led the congressional campaign for a peacetime national science foundation.

In 1950, Truman signed into law a compromise: the newly established National Science Foundation (NSF) would have a director appointed by the president, but would use a peer-review process to approve research money. Truman accepted the compromise because he had been persuaded by Kilgore that the training of scientists was a priority, and that the best institutions to undertake that function—namely the independent universities of the Ivy League—would accept only peer-reviewed grant money. The subsequent expansion of federal funding of science provided another natural experiment.

As noted above, as late as 1940, the federal and state governments between them funded only 23 percent of U.S. R&D, and almost no basic science, yet the federal government was soon funding most U.S. basic science and most R&D as well. The effects on the long-term rates of per capita GDP growth in the United States were *zero*. Since circa 1830, U.S. per capita GDP had been growing at around 2 percent per year, and after 1940 that rate did not rise. Likewise, after its governments began funding research during the 20th century, there was no upward trend in the long-term rate of per capita GDP growth in the UK.[62]

The State of Science Today

There can be no doubt that science today is in trouble. Consider the crisis of reproducibility: John Ioannidis at Stanford University has found, in the dramatic words of the title of one his papers, that "most published research findings are false"; Brian Nosek at the University of Virginia has reported that fewer than half of published studies in psychology can be reproduced.[63] Though some of the details in the Ioannidis and Nosek papers have been challenged,[64] no one disputes that science seems to have stumbled into a crisis of reproducibility. Why?

The answer is, of course, university scientists' need to publish in order to advance in their careers. Scientists in the public sector have an incentive, in the words of Paul Smaldino and Richard McElreath of the University of California, Merced, and the Max Planck Institute, Leipzig, to select "methods of analysis . . . to further publication instead of discovery." They report that *entire scientific disciplines*—despite isolated protests from whistleblowers—have, for over half a century, selected statistical methods precisely because they will yield publishable, though not necessarily true, results.[65]

The production of positive results facilitates publication. A 2012 paper by Daniele Fanelli (then at Edinburgh University) summarized its findings in the title: "Negative Results Are Disappearing from Most Disciplines and Countries." Using a sample of 5,000 published papers, Fanelli showed that there is a large and systematic increase in the percentage of studies reporting positive results in support of the researcher's original hypothesis.[66] Ioannidis moreover confirmed that the demand to publish and get funding is so strong that many studies are designed either to produce positive results or to produce experimental data that support the original research hypothesis, or both.[67]

The more a country adopts the "American model" for advancement, which is the model that has flourished since the war, with the NSF and the National Institutes of Health funding university science competitively using a peer-review process, the more that positive results are disproportionately published. Together with Ioannidis, Fanelli found that if an international team of scientists invites an American to participate in a study, the probability of a positive result doubles.[68] An American academic's advancement relies on maintaining funding for his research, and he is therefore loath to conclude that his funded hypothesis is not supported by the data. The epidemic of positive results, as Silas Nissen of the University of Copenhagen has noted (as illustrated in Figure 1.1), results in "the canonization of false facts."[69]

FIGURE 1.1

U.S. papers reporting support for the stated research hypothesis

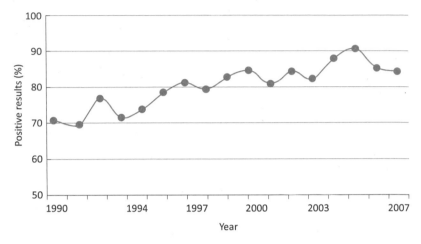

Source: D. Fanelli, "Negative Results Are Disappearing from Most Disciplines and Countries," *Scientometrics* 90 (2012): 891–904.

Note: (overall effect of interaction) Wald = 2.607, df = 3, P = 0.456.

Negative findings (i.e., those that do not confirm the original research hypothesis) are doubly unwelcome because journals find them unexciting and because funding agencies are reluctant to renew support for results that counter the funded hypothesis. The resulting "file-drawer effect" finds such work filed away until subsequent reanalyses and hypothesis adjustments can produce a "positive" result. This is also called "publication bias," which was defined (among many other related definitions) by Hans-Hermann Dubben in 2005 as "a well known phenomenon . . . in which positive results have a better chance of being published, are published earlier, and are published in journals with higher impact factors."[70]

The crisis of reproducibility is worse in disciplines that rely on statistics. In the so-called hard sciences of mathematics, physics, astronomy, and chemistry, individual phenomena can be isolated, which is why (to generalize) statistics are less important in the hard sciences than in the so-called soft sciences of psychology, sociology, economics, and other social sciences. In those soft sciences, phenomena can be deeply intertwined, so their effects need to be isolated by means of statistics. But statistics are easily abused, and Fanelli has shown that the rate of positive findings is fivefold greater in soft sciences than in hard ones. In plain language, the studies have been manipulated to appear to find something interesting.

Statistics can be abused in numerous ways. Among them are "p-hacking," in which datasets or hypotheses are adapted until a result reaches the bound for statistical significance,[71] and "data dredging," where enough variables are tested to ensure that at least one emerges as statistically significant. If, for example, 20 variables are studied—say the effects of calorie intake, blood pressure, blood glucose levels, exercise, and so on, on heart attacks—there is a 50-50 chance that one of the 20 will be significant at the .05 level because, as any statistician can tell you, a .05 probability is just a one in 20 chance. But in a dredged study, only the one significant variable is written up, which thus yields an apparently positive result.[72]

HARKing—"hypothesizing after the results are known"—can happen when a dredged result generates a so-called prior hypothesis that was never actually considered before the data were examined.[73] If, for example, low blood pressure emerged as significantly associated with heart attacks, a paper might begin: "Many metabolites at both high and low concentrations are associated with cardiovascular disease, so we hypothesized a similarly biphasic relationship for blood pressure." This hypothesis would produce a positive result, with no published mention of the many other potential predictors.

There are other ways to generate a positive result. The earth's surface temperature varies, and it is easy to pick start and end dates to show warming, cooling, or constancy. It then becomes easy to correlate those dates with a particular human activity and thus report an apparent cause and effect.

The sources of bias in science are myriad.[74] In addition to those already noted, they include small-study effects (the greater likelihood that studies with small numbers of observations will demonstrate marked differences relative to control groups); gray literature bias (writing or citing publications that do not appear in the mainstream peer-reviewed literature); the decline effect (when initial published findings are dramatic in their effects but emerge as much less dramatic after the study is reproduced by other researchers); citation bias (when researchers give undue credibility to certain publications by citing them preferentially); industry bias (the tendency of scientific publications to yield results that support the interests of their funders); government bias (the tendency to find results consistent with the politics of the funding source); and the U.S. effect (preferential publication of positive results). Fanelli and colleagues have found that though these problems are rampant, they are not the fundamental cause of science's distemper. That turns out to be a systematic problem with funding.

It was Daniel Sarewitz of Arizona State University who located the problem in the model of science funding that has dominated in the United States since the creation of the NSF and the expansion of the NIH after World War II. In this peer-review model, scientists are accountable (as Truman had anticipated) only to fellow scientists. Consequently, the biases described here can be harnessed to generate findings that apparently confirm the dominant paradigm rather than reflecting nature: in Sarewitz's words, it is "technology that keeps science honest," but the whole point of government funding of science, and of peer review, is to uncouple scientists from the demands of technology.[75]

These problems were anticipated not only by Truman but also by his successor. In his famous "military-industrial complex" Farewell Address of 1961, President Dwight D. Eisenhower lamented the effects on universities of the flood of federal money Truman had unleashed, and the possibility that scientists would use it to influence national policy:

A steadily increasing share [of research] is conducted for, or at the direction of, the federal government [so] the free university, historically the fountainhead of free ideas and scientific discovery, has experienced a revolution in the conduct of research. Partly because of the huge costs involved, a government contract becomes virtually a substitute for intellectual curiosity. . . . The prospect of domination of the nation's scholars by federal employment, project allocations and the power of money is ever present—and is gravely to be regarded. Yet in holding scientific research and discovery in respect, as we should, we must also be alert to the equal and opposite danger that public policy could itself become captive of a scientific-technological elite.[76]

As late as the 1950s, American universities had supposed their role was to pursue scholarship and educate the young. They had not realized that their role also included rescuing the market from its manifest failure by turning themselves into government-funded scientific research labs that did a bit of teaching on the side and eschewed any of that speaking-truth-to-power stuff, at least if that power distributed research grants. But they were soon put right, and Fred Stone of the NIH, for example, noted that during the 1950s "it wasn't anything to travel 200,000 miles a year" to coax grant applications from the universities and to disburse research gold if they only filled out the forms correctly (in those days, most grant applications were funded).[77] Before long, the universities wouldn't need the coaxing, and the grant awards would get much more selective.

Today in the United States, the government funds half of all pure science, a quarter is funded by charitable foundations, and the remaining quarter is funded by industry. But much of the charitable foundation and industrial money for pure science is directed to organizations outside the universities, so the source of U.S. university funding is largely the government. And the consequence is that pure science has changed from being something companies and foundations funded in their search for truth, into something that governments fund by peer review within the universities.

The public supposes that peer review provides an imprimatur of credibility to research results, but it is possible for it to do no such thing. Credibility depends on the honesty of the researchers, which can be assessed only after third parties attempt to reproduce their results. Peer review, therefore, can at best only screen for plausibility. At worst, it is a vehicle for imposing groupthink. And groupthink is the curse of public science today, because fashionable groups feed on fashionable funds. This is ultimately why we have seen an increasing number of articles retracted after publication.[78]

Moreover, because the distribution of scientific ability is skewed according to the power law,[79] the more money that is awarded to a finite research community, the worse the quality of the science becomes. This should especially be true in a field such as climate science, which has very few PhD holders but receives a tremendous amount of research money.

The Politicization of Science

In 1605, Francis Bacon described science as a four-stage process of (1) observation, (2) induction, (3) deduction, and (4) experimentation, and Karl Popper, in his 1959 book *The Logic of Scientific Discovery*, argued that if a theory was not falsifiable then it was not scientific. Most people suppose that that pretty much sums up "science," but at least three philosophers—Michel Foucault, Paul Feyerabend, and Imre Lakatos—have argued that science in practice is much less rational or objective, and much more biased, than is generally understood. One problem (to encapsulate a complex debate) is that scientists must choose what to observe.

Bacon supposed that scientists simply observed the world, but the world offers such a mass of data that scientists have to select their observations (e.g., does the color of your socks matter when you are sequencing a stretch of DNA?); and in choosing, they may predetermine their findings. As Einstein said, "Your theory determines what you observe." If, for example, you believe that fat (not sugar) causes heart attacks, then—diets being so various—it is only a matter of time and application before you can find some dietary fat in some population that correlates with heart attacks, thus "proving" your theory.

But is it your theory? Or is it your funders' theory? Or is it the theory of the editors of the top journals, and of the referees at the granting agencies, and of the members of your promotions and tenure committee? If university science is going to be rewarded by peer review, then inevitably it becomes political, as long as there is a substantial stream of publicly appropriated financial support, and researchers will do well to pick the observations and findings that resonate with those who control their careers and incomes, as Marc Edwards discovered.

Marc Edwards, a professor of civil engineering at Virginia Tech, uncovered the lead contamination in the Flint, Michigan, water supply. In the *Chronicle of Higher Education* he told a story that closely parallels Ibsen's 1882 play *An Enemy of the People*, in which a man who tells an unpopular truth is vilified for it. By challenging the authorities' reassurances about the quality of the water, Edwards risked alienating potential sources of research funding at his university. For this he was not thanked by his peers and colleagues.[80]

Edwards condemned the reluctance of scientists to challenge authority for fear of losing funding; in an interview he said that the systems built to support scientists do not reward moral courage, and that the university pipeline contains toxins of its own, which if ignored will corrode public faith in science:

> I am very concerned about the culture of academia in this country and the perverse incentives that are given to young faculty. The pressures to get funding are just extraordinary. . . . This is something that I'm upset about deeply. I've kind of dedicated my career to try to raise awareness about this. I'm losing a lot of friends.[81]

Further,

> I know the culture of academia. You are your funding network as a professor. You can destroy that network that took you 25 years to build with one word. I've done it. When was the last time you heard anyone in academia publicly criticize a funding agency, no matter how outrageous their behavior? We just don't do these things.[82]

Universities worry about contrarian research because they need the overhead money that is included in federal grants. Government agencies not only support the direct costs of a research project but also add a premium of around 50 percent, which contributes to maintaining the infrastructure of the university and renders the whole institution, from the library to the department of romance languages to the president's office, dependent on government research grants. Thus the entire university has an interest in government funding and in the government's goals. And the sums of money are considerable.

Most faculty are paid for only nine or ten months a year. When offered a salary of $100,000, that sometimes means $75,000, unless someone else—almost always the federal government—pays. "Summer salary" is a part of most grant proposals.

University faculty spend much time writing funding proposals, and often much of the research is performed by postdoctoral researchers on short-term contracts. Each postdoc will cost about $80,000 per year, while their research assistants, usually graduate students, cost around $30,000 apiece.

Consider a proposal to study the hypothesis that global warming will increase weather-related morbidity and mortality. Table 1.2 shows an estimated budget, with summer salary for one faculty member, two postdoctoral researchers, and three graduate research assistants, if applied for today by the University of Virginia (UVA).

TABLE 1.2
Estimated UVA department budget

Position		Year 1, inflation-adjusted three-year total (dollars)
Principal investigator (faculty)		
Summer salary	33,000	103,942
Postdoctoral researchers (2)	160,000	504,400
Graduate research assistants (3)	90,000	283,725
Total wages		788,125
Fringe benefits @ 27.7 percent		218,311
Total salary and benefits	1,006,436	
Subtotal		1,085,247
UVA indirect costs (overhead) 54.0 percent	586,033	
Total		1,671,280

The amount of published research necessary for a faculty member to achieve tenure would probably entail about three federal contracts of this magnitude, so, about $5 million in total, with the university banking about $1.8 million in overhead.[83] The nonscience departments that cannot recover their expenses in tuition and fees get much of this, and the big research universities are dependencies. Threatening the consensus in science, therefore, if it also threatens the chances of winning grants, will threaten an institution's financial health, which inevitably arouses concern.[84]

There is a certain irony in state funding being the corruptor of university integrity. On December 10, 1819, Thomas Jefferson, who that year had founded the University of Virginia, wrote to John Adams describing the benefits of an educated citizenry:

> Their minds were to be informed, by education, what is right & what wrong, to be encouraged in habits of virtue, & deterred from those of vice by the dread of punishments, proportioned indeed, but irremissible; in all cases to follow truth as the only safe guide, & to eschew error which bewilder us in one false consequence after another in endless succession. These are the inculcations necessary to render the people a sure basis for the structure of order & good government, but this would have been an operation of a generation or two at least, within which period would have succeeded many Neros and Commoduses, who could have quashed the whole process.[85]

Jefferson's motive in creating the university was to create a secular institution free of the religious influence that had so offended him when he was studying at the College of William and Mary. Unfortunately, universities today do not live up to his ideals, and they betray free inquiry and free speech in ways he did not fully anticipate.[86]

Tenure—which is meant to protect free speech—turns out to be just another facet of the problem. Tenure binds a group of scholars together indefinitely, encouraging them to maximize their adherence to the group. Scientists do not speak out, as Edwards noted, because they fear losing their research funding and their close colleagues.

In this book we have assembled ten examples—ranging from dietary advice and opioids to copper and uranium mining to climate change to the war on drugs to the regulation of carcinogens and other chemicals— of how the modern scientific enterprise can be used, abused, or misused in the service of massive regulation or to deprive people of property rights. Everyone knows that the government has the technology to read our most private communications, which can and will be used to constrain liberty, but few people can see that science, when distorted by the government, will spontaneously generate intellectual and cultural abuses that can undermine our freedom.

In this book we will see how politicized claques can authoritatively prescribe deadly diets with impunity, with separate—but remarkably similar—stories on dietary fat and salt. In both cases, federal guidelines are causing more harm than good. The rest of this book details other disasters, largely unfolding unnoticed by a public that supposes science is a value-free enterprise, that it always tells the truth, and that it deserves public support.

Besides incentive-based bias, there is another way in which government can influence policy or take away property rights, one that might be termed "abuse of scientific authority." This can occur when a federal agency or federally chartered entity, such as the National Academy of Sciences, decides to speak out on an issue of scientific controversy. Other federal agencies can engage in similar authority badgering.

What do we do about the failings of science? Can we reform science while keeping government funding, or should we revert to the pre-1940 world where it was free of government, where it was kept honest by its link to technology and markets, and where it rode high in public esteem and trust? We address these questions here by looking carefully at the ways in which the federal government has impinged on the scientific process.

LARDING THE SCIENCE

The Dietary Fat Fiasco

The U.S. federal government's agencies have been issuing dietary advice for more than a century, but before 1977 they limited themselves largely to addressing malnutrition among the poor. The first two-thirds of the 20th century had witnessed the early triumph of nutrition research when, in a world still concerned with malnourishment, the discipline had helped oversee the discovery of vitamins and the elaboration of the basic principles of metabolic biochemistry. So in 1968, when the Senate established the Select Committee on Nutrition and Human Needs led by Sen. George McGovern (D-SD), it focused initially on the problems of undernutrition. But a decade later, on January 14, 1977, when it published its *Dietary Goals for the United States*, the committee launched an attack on the apparent problems of *over*consumption. In his introduction, McGovern wrote:

> This is the first comprehensive statement by any branch of the
> federal government on risk factors in the American diet.
> Too much fat . . . [is] . . . linked directly to heart disease, cancer,
> obesity and stroke.

... Six out of the ten leading causes of death in the United States [heart disease, cancer, vascular disease, diabetes, arteriosclerosis and cirrhosis of the liver] have been linked to our diet.[1]

The committee reported unanimously, and in his foreword Sen. Charles Percy (R-IL) wrote, "Without government . . . commitment to good nutrition, the American people will continue to eat themselves to poor health."[2] Consequently, the committee explained, "We as a government . . . have an obligation to provide practical guides to the individual consumer as well as to set national dietary goals for the country." Accordingly, Americans were urged to:

1. Increase carbohydrate consumption to account for 55 to 60 percent of energy (caloric) intake.
2. Reduce overall fat consumption from approximately 40 to 30 percent of energy intake.
3. Reduce saturated fat consumption to account for about 10 percent of total energy intake.
4. Reduce cholesterol consumption to about 300 mg a day.[3]

The committee had, in short, officially launched the anti-fat, pro-carbohydrate campaign that was to dominate the world of nutrition until recently and which still reigns in official circles.

Why Did the Senate Select Committee Launch an Attack on Fats?

The problem was heart attacks. The heart attack epidemic had seemingly come out of nowhere and by 1968, at its height, accounted for more than one-third (37 percent) of all deaths in the United States. By contrast, all cancers accounted for only one-sixth (17 percent) of all deaths, and strokes only one-tenth (10 percent). Accidents, at 6 percent, were the fourth most common cause of death.[4] The sudden epidemic of heart attacks was profoundly alarming, especially as it seemed to target otherwise healthy people at the peak of their performance.

Some physicians argued that the epidemic was illusory, the result of better diagnosis and an aging population; yet that argument, though not trivial, was to be disproved. In 1966, for example, Leon Michaels, a Canadian physician, showed that the lack of evidence of heart attacks before the 20th century was indeed evidence of their absence: on comparing the characteristic chest pain of heart attacks with the characteristic pain and symptoms of migraine and gout, he showed that, whereas frequent descriptions of migraine and gout can be found in medical texts from all eras, stretching back to Greek and Roman times, angina and heart attacks started to be described with any frequency only in the 20th century.[5] He thus concluded that the death rates from those diseases had increased up to two hundredfold between 1901 and 1962, and although the rate of increase cannot be known with certainty, it is now accepted that the rise in the incidence of heart attacks during the 20th century was real (see Figure 2.1).

Faced with such an epidemic, some commentators argued that it was surely reasonable for the U.S. government to address it. In 1974, for example, Marc LaLonde, Canada's minister of national health and welfare, had published a report intended to prescribe the diet of the Canadian people, in which he suggested that Canadians should eat less fat—in particular less saturated fat and less cholesterol—and more carbohydrates.[6] Some commentators argued it was surely not unreasonable for the U.S. Senate Select Committee on Nutrition and Human Needs also to take a position.

It might indeed not have been unreasonable—had the committee's position been a wise one. Yet the federal government may be institutionally incapable of providing wise dietary advice.

Why the Demonization of Fat?

The senators on the Select Committee were not, of course, nutritionists, so they took their lead from the scientists, and from one scientist in particular, Ancel Keys (1904–2004). Keys, who was the professor of physiology at the University of Minnesota, launched the modern dietary era in 1953 with the publication of "Atherosclerosis: A Problem in Newer Public Health."[7]

By 1953, the Food and Agriculture Organization of the United Nations (FAO) had collected dietary information on 22 countries. Keys selected data from six countries (Australia, Canada, England

FIGURE 2.1

Death rates per 100,000 people from atherosclerotic heart disease and stroke

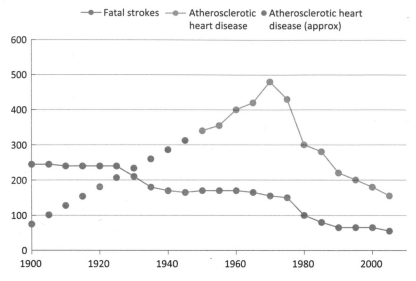

Sources: D. T. Lackland et al., "Factors Influencing the Decline in Stroke Mortality: A Statement from the American Heart Association/American Stroke Association," *Stroke* 45 (December 5, 2013): 315–353, doi:10.1161/01.str.0000437068.30550; Leon Michaels, "Aetiology of Coronary Heart Disease: An Historical Approach," *Heart* 28 (March 1966): 258–264, doi:10.1136/hrt.28.2.258.

Note: The data in solid lines come from the joint 2013 statement of the American Heart and Stroke Associations. Data on atherosclerotic heart disease were not collected before 1950, so the dotted lines are extrapolated from the estimates in the Michaels article.

and Wales, Italy, Japan, and the United States) to generate a graph in which he showed that as the percentage of fat in the diet rose, so the death rates from heart disease of males aged 55–59 years rose accordingly: from 0.5 per thousand in Japan (where the typical diet included 7 percent fat) to just under 7 per thousand in the United States (with a diet of 40 percent fat).

In 1953, as today, heart disease was understood essentially as the consequence of atherosclerosis. But whereas today we understand atherosclerosis to be an inflammation of the arteries, in 1953 it was seen more as a hardening (technically an arteriosclerosis) caused by the deposition of cholesterol within the arterial walls. Keys proposed that as people ate too much fat (in particular too much cholesterol), so the blood vessels silted up with cholesterol, leading the heart to become diseased as its narrowed arteries failed to provide enough oxygen or nourishment. That process in turn would lead to a heart attack or myocardial infarction as the blood clotted over the cholesterol-filled artery and thus killed the patient. Here was Keys's model:

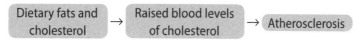

Initial Criticism of Keys's Model

It is often forgotten that, from the beginning, Keys's model was heavily criticized. In 1955, for example, the World Health Organization (WHO) convened a small seminar of international experts who proceeded comprehensively to demolish it. (One of his colleagues remembered the WHO meeting as "the pivotal moment in Keys's life. He got up from being knocked around and said, 'I'll show these guys.'"[8]) Among the critics were Jacob Yerushalmy and Herman Hilleboe, from the University of California, Berkeley, and the New York State Commission of Health, respectively, who attacked the model both at the seminar and shortly afterward in a paper.[9]

At the 1955 seminar Keys had claimed, "There is a remarkable relationship between the death rate from degenerative heart disease and the proportion of fat calories in the national diet," which was supportable. But then he claimed, "No other variable . . . shows anything like such a consistent relationship," which was not supportable.[10] When Yerushalmy and Hilleboe reexamined the Food and Agriculture Organization data, they found the association between animal protein and heart disease to be stronger than between fat and heart disease. And the association between the total consumption of calories and heart disease was stronger yet. In addition, the strongest determinant of calorie and meat intake seemed to be GDP per capita.

On the other hand, consumption of vegetables seemed to protect against heart disease, even though vegetables contain fat, protein, and carbohydrates. To further complicate matters, it appeared that the greater the consumption of animal fat and protein, the lower the death rates from every condition except heart disease. So, the model that best accounted for the empirical facts was:

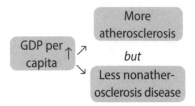

As Yerushalmy and Hilleboe pointed out at the 1955 WHO seminar, and as they expanded in their 1957 paper, the data thus suggested that the citizens of poor countries (who ate primarily vegetables, including starchy vegetables such as maize or corn, rice, and potatoes) did not often die of heart disease (but were vulnerable to other diseases); the citizens of rich countries (who ate a lot of meat, which includes much fat) died largely of heart disease (but were protected from other causes of death).

And to confirm that GDP per capita seemed central to the development of heart attacks, Yerushalmy and Hilleboe noted that atherosclerosis remitted when people reverted to a pre-Western lifestyle: during World War II and its aftermath, many parts of Europe had been reduced to meager diets, and those parts of Europe saw their rates of heart disease fall; but when normal food supplies were restored, heart disease returned. Heart disease, in short, seemed to be a consequence of a Western diet (which was in turn a consequence of Western wealth), but no single element of that diet could be identified as especially responsible.

The discussions at the 1955 WHO seminar were prescient, because the delegates saw ischemic heart disease not as the consequence of a single cause (cholesterol) but as the consequence of many, complex, still-to-be-elucidated causes. More of those possible causes were soon identified, and when in 1957 professor of physiology John Yudkin of the University of London reexamined the FAO data, he found "a moderate but by no means excellent relationship between fat consumption and coronary mortality.... A better relationship turned out to exist between sugar consumption and coronary mortality in a variety of countries."[11]

Yudkin therefore proposed that sugar, not fat, was bad for the heart, and he wrote of "good nutritious foods like meat and cheese and milk."[12]

FIGURE 2.2

Male peptic ulcer death rates and per capita cigarette consumption

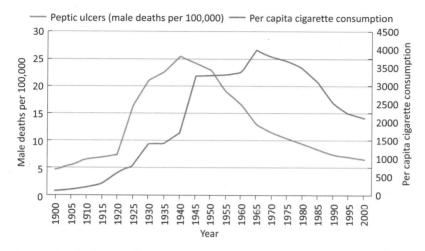

Source: Alexander Mercer, *Infections, Chronic Disease, and the Epidemiological Transition: A New Perspective* (Rochester, NY: University of Rochester Press, 2014), p. 184. The data on cigarette consumption come from Centers for Disease Control and Prevention, *Achievements in Public Health, 1900–1999: Tobacco Use, United States, 1900–1999, Morbidity and Mortality Weekly Report* 48 (Washington: CDC, 1999): 986–993.

Also in 1957, Edward Ahrens (1915–2000) of the Rockefeller Institute of Medical Research—the doyen of fat biochemistry—who had long recognized carbohydrates as cardiac killers, protested that "unproved hypotheses are enthusiastically proclaimed as facts."[13] (Ahrens later greeted the Senate Select Committee on Nutrition and Human Needs' recommendations as treating people as if they were "a homogenous group of Sprague-Dawley rats.")[14] George V. Mann, a Vanderbilt University biochemist, was yet another researcher who had shown that Keys's demonization of animal fats did not fit the cardiac facts.[15]

Meanwhile, in 1970, Richard Doll, the epidemiologist who had earlier reported that cigarettes caused lung cancer, found: "It is cigarette smoking . . . which is implicated in the aetiology and manifestation of myocardial infarction."[16] Figure 2.2 illustrates that one nondietary phenomenon that tracks the incidence of atherosclerotic heart deaths is cigarette smoking.

FIGURE 2.3
Proposed causes of atherosclerosis, 1970s

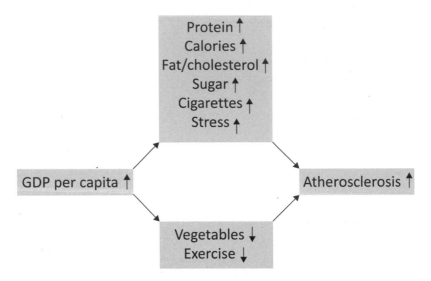

Source: These data were extracted by the author from the contemporary literature.

The list of possible causes did stop there; additional causes, including a lack of exercise and excessive stress, were soon identified. By the 1970s, therefore, the criticisms of the 1955 WHO seminarians had been vindicated, and the only sure model compatible with the data was that found in Figure 2.3, but which of the many possible factors was primarily responsible for the epidemic of heart attacks could not be isolated.

So why in 1977 did the Senate committee back Keys's dietary fat hypothesis over all the other possible causes of atherosclerosis? Well, Keys had generated a model: whereas no one could easily suggest how calories, protein, cigarettes, sugar, a lack of exercise, or a lack of vegetables could provoke atherosclerosis, Keys *could* suggest how fat could do it.

Keys's First Model for Atherosclerosis

Keys first noted that heart attacks were caused by atherosclerosis, that atherosclerotic plaques were full of cholesterol, that patients who suffered heart attacks had elevated blood levels of cholesterol, and that rabbits that were fed high levels of cholesterol develop atherosclerosis.[17] He thus proposed:

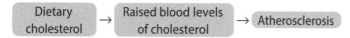

By 1955, though, Keys had realized that dietary cholesterol is not a human danger: our livers synthesize most of our cholesterol, and when we ingest it in our food, our livers simply reduce their creation of cholesterol.[18] This is not true of all animals, particularly not of herbivores such as rabbits, which—because plants are low in cholesterol—do not normally handle significant amounts of it. So herbivores, when fed high cholesterol in the laboratory, respond with raised cholesterol in the blood: their livers don't know any better. But we are omnivores, our livers are not naive, and when fed high cholesterol we do not respond with raised levels of blood cholesterol. For us:

Dietary cholesterol ≠ Circulating cholesterol

Asymmetrical Science

In 1955, within two years of proposing the dietary cholesterol hypothesis, Keys had abandoned it. But for another 60 years the U.S. federal government continued to warn against consuming cholesterol-rich foods. It was only in 2015 that its Dietary Guidelines Advisory Committee classified high-cholesterol foods such as eggs, shrimp, and lobster as safe to eat: "cholesterol is not a nutrient of concern for overconsumption."[19]

This 60-year delay shows how asymmetrical the official science of nutrition can be: a federal agency can label a foodstuff dangerous based on a suggestion, yet demand the most rigorous proof before reversing its advice. Harvard professor of epidemiology and nutrition Walter Willett, commenting on the asymmetry in a related area of government nutrition advice, described it as "Scandalous. They say 'You really need a high level of proof to change the recommendations,' which is ironic, because they never had a high level of proof to set them."[20]

And it was Keys himself who championed asymmetry in dietary advice when he wrote in 1957 that nobody had "adequate evidence to state that there is not a causal relationship between dietary fat and the tendency to develop atherosclerosis in man" (i.e., he could condemn fat on the basis of a hypothesis only, yet it could only be classified as safe after exhaustive study).[21] So Keys not only launched the cholesterol/fat paradigm on inadequate evidence, he also biased the debate in its favor.

Further, it may seem safer to advise abstention from a particular food than to declare one to be safe. Yet abstention may itself be dangerous, because abstention can have unintended consequences: other foodstuffs must be consumed instead. The Select Committee's major scientific adviser was Mark Hegsted of Harvard, who wrote in his foreword to the *Dietary Goals* of 1977:

> The question to be asked, therefore, is not why we should change our diet but why not? What are the risks associated with eating less meat, less fat, less saturated fat, less cholesterol, less sugar, less salts, and more fruits, vegetables, unsaturated fat and cereal products—especially whole grain cereals?[22]

The answer to this question—namely, that it would lead to the eating of more carbohydrates and more trans fats, both of which really are dangerous—would soon emerge. (Trans fats are chemically synthesized unsaturated fatty acids associated with increased risk of coronary heart disease.)

Meanwhile, the 60 years of official misinformation has taken its toll: a 2015 survey by Credit Suisse Research Institute (a social research charity) found that 54 percent of doctors falsely believed eating cholesterol-rich food raises blood levels of cholesterol and damages the heart. In the words of the survey, "This is a clear example of the level of misinformation that exists among doctors."[23]

Keys's Second Model

Keys always saw dietary cholesterol as only one of two problematic factors, and by 1955 he was presenting his second set of facts. He noted that eating fat, especially saturated or animal fat, raised the blood levels of cholesterol, from which he generated his second model:

| Dietary saturated fats | → | Raised blood levels of cholesterol | → | Atherosclerosis |

This, of course, was the model the Select Committee was to eventually endorse in 1977. But as we have seen, on its being presented to the WHO delegates back in 1955, they had immediately been skeptical,

> The evidence is circumstantial . . . no conclusions of etiological [causative] relationships should be attempted unless the factor is found to be related to the disease by evidence from entirely different types of investigations.[24]

To counter this skepticism, Keys had launched his famous *Seven Countries*, published in 1970, in which he personally examined what people were eating in Finland, Greece, Italy, Japan, the Netherlands, the United States, and Yugoslavia. He refined his surveys to look at saturated (i.e., animal) fats rather than total fats in the diet, and yet again he found a strong association between (saturated) fat ingestion and deaths from heart disease.[25]

But Keys's *Seven Countries* wasn't the "entirely different type of investigation" for which the WHO experts had called, since it still generated data that they considered "circumstantial."[26] Indeed, when one of Keys's colleagues, Alessandro Menotti, recharacterized the foodstuffs in *Seven Countries,* he found "sweets" (sugar-rich products, cakes, and other confectioneries) in the diet correlated more strongly with coronary mortality than did "animal food" (butter, meat, eggs, margarine, lard, milk, and cheese). Even Keys's own program of work, therefore, suggested it might be carbohydrates, not fats, that killed people.[27]

Bad Science

The Select Committee's gravest offense was not merely going beyond the verifiable science, but flouting it. There is in epidemiology a "hierarchy of evidence": some data are recognized to be more credible than others, and in particular randomized controlled trials (i.e., experiments) are recognized to produce harder data than observations (which may report only associations, and which may, in turn, mislead).

By 1977, when the Select Committee published its report, no fewer than six randomized trials had been performed on a total of 2,467 males (of whom 423 died from heart problems during the trials) in which their total and saturated fat intake was reduced by placing them on low-fat diets. As predicted by Keys's hypothesis, for the subjects on reduced dietary fat, the circulating blood levels of cholesterol declined. But contrary to the hypothesis, *their mortality rates did not fall.* In their devastating review of the six trials, Zoë Harcombe and her colleagues in Wales and Kansas City concluded, "Dietary recommendations were introduced for 220 million U.S. citizens . . . in the absence of supporting evidence from randomized controlled trials."[28]

That lack of hard evidence was recognized at the time, and in 1977 the American Medical Association responded to *Dietary Goals* with the statement:

> The evidence for assuming that benefits [are] to be derived from the adoption of such universal dietary goals . . . is not conclusive, and there is potential for harmful effects.[29]

Though the committee had acknowledged the incomplete state of the science of the day, it had also written with approval:

> Marc LaLonde, Canada's Minister of National Health and Welfare, said: "Even such a simple question as whether one should severely limit his consumption of butter and eggs can be a matter of endless scientific debate . . . [so] it would be easy for health educators and promoters to sit on their hands. . . . But many of Canada's health problems are sufficiently pressing that action has to be taken even if all the scientific evidence is not in."[30]

When challenged on the incompleteness of the science, Senator McGovern said, "Senators do not have the luxury that the research scientist does of waiting until every last shred of evidence is in," which is the opposite of the truth: research scientists are at leisure—and are perhaps even obligated—to explore every possible hypothesis, but senators should not issue advice *until* every last shred of evidence is in, because they may otherwise issue misleading or even dangerous advice.[31] As they did in 1977.

We see here a second reason why official dietary advice may be institutionally biased. Officialdom may be under pressure to issue it prematurely, sometimes even by decades. Moreover, this advice may be based on models rather than on hard facts.

Why Did the Randomized Controlled Clinical Trials Not Confirm the Dietary Saturated Fat Model?

The Keys model was two-staged:

Dietary saturated fat $\rightarrow$ Raised levels of blood cholesterol

and

Raised levels of blood cholesterol $\rightarrow$ Atherosclerosis

Confusingly, both stages of the model were true. Mark Hegsted, the head of the Department of Nutrition at Harvard and the Select Committee's major adviser, had shown that when humans ate saturated fat, their circulating blood levels of cholesterol did indeed rise. Equally, two future Nobel laureates from Texas, Michael Brown and Joseph Goldstein, showed that—in certain inherited diseases of metabolism—high blood levels of cholesterol can indeed cause heart attacks. So, the committee put two and two together and supposed:

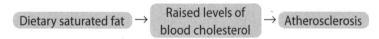

Dietary saturated fat $\rightarrow$ Raised levels of blood cholesterol $\rightarrow$ Atherosclerosis

thus reminding us of H. L. Mencken's aphorism that for every complex problem there is a solution that is clear, simple, and wrong. As we saw above, when no fewer than six randomized controlled clinical trials had tested the complete model by withdrawing saturated fat from the diet of vulnerable men, their blood levels of cholesterol fell but their heart death rates did not. Why not?

There are at least three different types of circulating blood cholesterol. One type, so-called HDL (high-density lipoprotein) is positively healthful, as it draws cholesterol out of the arteries. Another type, lLDL (large low-density lipoprotein) is largely neutral. A third type, sLDL (small low-density lipoprotein) is the one that kills, as it (and its oxidized forms) tend to lodge in the arteries and precipitate the inflammation we know as atherosclerosis. Moreover, saturated fat in the diet tends to raise the circulating levels of the essentially neutral lLDLs, while carbohydrate in the diet tends to raise the circulating levels of the dangerous sLDLs. Therefore, it is carbohydrate, not fat, in the diet that raises the dangerous type of cholesterol, although the mechanisms remain to be fully elucidated.

So it is no surprise that the six randomized controlled clinical trials failed to find a decline in the rate of heart death rates following the reduction of saturated fat in the diet of vulnerable men. This is because the compensatory rise in carbohydrate intake was dangerous. To Mark Hegsted's question in the foreword to *Dietary Goals*, "What are the risks associated with eating less meat, less fat, less saturated fat, less cholesterol?," we can now reply that if, instead, people were to follow his advice and eat more carbohydrates and more trans fats in compensation, they would risk precipitating early death from atherosclerosis. Irony of ironies.

The American People Were Dutiful

Nonetheless the American people did as they were advised—not just by the government, but also by the mass media, which reinforced the government's message.[32] In a 1986 survey conducted jointly by the National Heart, Lung, and Blood Institute and the Food and Drug Administration, 72 percent of adult respondents "believed that reducing high blood levels of cholesterol would have a large effect on heart disease."[33] Between 1960 and 2000, their per capita consumption of saturated fatty acids fell from 55 to 46 grams per day, and their per capita consumption of cholesterol fell from 465 to 410 milligrams per day. Meanwhile, their per capita consumption of carbohydrates rose from 380 to 510 grams per day, and consumption of fiber rose from 18 to 26 grams per day (Figure 2.4).[34]

Americans also increased their intake of unsaturated vegetable-derived fats chemically modified as trans fats (for example, replacing butter with margarine) because it was believed that saturated fat was

FIGURE 2.4

Obesity and the consumption of different foods in America, 1960–2000

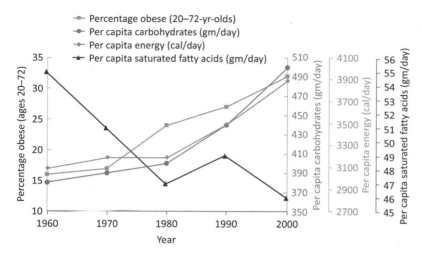

Source: The data come from Shi-Sheng Zhou et al., "B-Vitamin Consumption and the Prevalence of Diabetes and Obesity among U.S. Adults: Population Based Ecological Study," *BMC Public Health* 10 (2010): 746, https://doi.org/10.1186/1471-2458-10-746.

more dangerous than unsaturated fat. So whereas in 1911 per capita consumption was 19 pounds of butter a year and 1 pound of margarine, by 1976 butter consumption had fallen to 4 pounds a year and margarine's had risen to 12.[35] Yet we now know that trans fats lower healthful HDLs, raise the dangerous sLDLs, *and* are inflammatory. Consequently, in the words of a recent authoritative review, they "contribute significantly to increased risk of coronary heart disease events."[36]

The Select Committee's demonization of saturated animal fat—which led the American people to increase their consumption of trans fats—can only have damaged the health of those who followed its advice.

Keys and the Select Committee Vindicated?

Nonetheless, in following the Select Committee's advice, an odd thing happened to the American people: their rates of heart disease and stroke fell, really quite dramatically (see Figure 2.1).

Before invoking Figure 2.1 as vindication of the Select Committee's advice, consider Figure 2.2, which illustrates the course of another 20th-century epidemic, that of peptic ulcers. It is often forgotten, but the 20th century witnessed an epidemic of peptic ulcers that mirrored that of heart attacks. Since no one has suggested that saturated fat causes peptic ulcers, we should be careful not to confuse correlation with causation: a disease may reflect the incidence of heart attacks without that incidence proving that saturated fat is responsible.

Moreover, the stroke data in Figure 2.1 are incompatible with the Keys hypothesis. Strokes are caused by the same atherosclerotic process as heart attacks, yet their epidemiology looks different: namely, they peak before the 20th century, and during the first half of the 20th century they appear to have declined even in the face of the high-fat Western diet. During the second half of the 20th century they continued to decline, even as the Western diet became lower in fat and higher in carbohydrates. Figure 2.1, indeed, shows that variations in dietary composition *cannot* explain the epidemiology of strokes. (The data in Figure 2.1 relate to deaths, not to disease incidence; and although it is possible that changes in therapies influenced the long-term trends, those changes would not have been sufficient to generate such different trend lines.)

Strokes and the Metabolic Syndrome

So, how can we explain the epidemiology of strokes? The World Health Organization reports that in 2015, the latest year for which we have data, the two most common causes of death among low-income economies were chest infections and diarrhea (i.e., bacterial and viral infections of the lungs and guts). Stroke was third.[37]

Strokes are a disease of poverty, a connection that British epidemiologist David Barker explained when he showed, unexpectedly, a correlation between maternal mortality rates and strokes.[38] In his words:

If you want to know how much . . . stroke there is in any city, any town, any rural village, do not count the hamburger outlets, the tobacconists, the playgrounds. Ask instead how many mothers died in childbirth several years ago.[39]

Maternal mortality rates are highest where mothers are poor and ill-fed, and where in consequence their fetuses are malnourished. A malnourished fetus is a fetus under metabolic pressure, so it must make a choice: which organs will it protect? Will it allow all its organs to be malnourished equally, or will it protect some organs at the expense of others? It appears that a malnourished fetus chooses to protect its brain (these responses are little different from an adult's: when adult mammals are starved, most of their organs shrink, the exceptions being the brain and, in the case of male mice, the testicles).[40]

So, when starved, the fetus will deprive its other organs of nourishment, and as a result grow into a short adult; and to help achieve that smallness, it will induce its muscles and other major organs to become resistant to insulin. The role of insulin is to direct the glucose we absorb from our food into our muscles and other major organs, so when we become insulin-resistant, their glucose uptake is suppressed, thus sparing it for the brain.[41] And, unexpectedly perhaps, insulin resistance contributes to atherosclerosis.

Malnourished, poor mothers, therefore, produce children who are prone to developing strokes. But well-nourished, rich mothers do not. Hence the slope of the line in Figure 2.1: as mothers in the West have grown increasingly well-nourished, so their babies have become less prone to developing strokes.

The Incidence of Heart Attacks in the 20th Century: What We Know

From the divergence between the two lines in Figure 2.1, we can see that the causes of strokes and heart attacks must be different. Although both are caused by atherosclerosis, this condition must affect the arteries of the brain and heart in subtly different ways. Therefore, all we can currently state with certainty is that atherosclerosis is an inflammation of the arteries, of which there are many possible causes, including the insulin resistance of the metabolic syndrome and raised levels of sLDL, but also including smoking, stress, hypertension, diabetes, and aging, as well as a range of inflammatory diseases including arthritis, lupus, chronic infections, and inflammations of unknown cause. Another inflammatory disease that can apparently cause heart attacks is peptic ulceration, which is caused primarily by infection with *Helicobacter pylori*, because there is an association between *H. pylori* infection and atherosclerosis.[42] Importantly, therefore, we still cannot know with certainty what accounted for the epidemiology of heart deaths in the 20th century.

Shattering the Cholesterol Story

Fifty years after Keys captured the Senate Select Committee's imagination, the failings of the cholesterol hypothesis had become so obvious that in 2007 Gary Taubes, a science journalist, could publish a book *Good Calories, Bad Calories*, which became a bestseller. In the book he claimed that carbohydrates in general, and sugar in particular, are dietary hazards and that natural fats are healthful.[43]

There was already a long scientific tradition led by such professors as Yudkin, Mann, and Ahrens arguing that carbohydrates and sugar, not fats, are the cardiac killers. In addition, Taubes had a populist predecessor, Robert Atkins (1930–2003), who in 1972 had published *Dr. Atkins' Diet Revolution*. Atkins was a New York cardiologist who found that, for weight loss, a diet low in carbohydrates and high in fat and protein worked.[44]

While Yudkin, Mann, and Ahrens had focused on sugar and carbohydrates as the cause of *heart attacks*, Atkins had focused on sugar and carbohydrates as the cause of *obesity*. And Taubes continued in Atkins's wake, noting that the epidemic of obesity (and type 2 diabetes) accelerated when fat in food was being replaced by carbohydrate (see Figure 2.4). Thus, between 1960 and 2000, the incidence of obesity more than doubled, from 13.4 percent of the population to 30.9 percent, while the incidence of type 2 diabetes rose even more markedly, from 2.6 percent to 6 percent of the population. Since 2000 it has continued to rise to nearly 10 percent today.[45]

Yet we should eschew easy myth-making. Jennie Brand-Miller at the University of Sydney, Australia, has shown that in Australia and the

United Kingdom sugar consumption fell between 1980 and 2003 even as obesity rose, which suggests that obesity cannot be attributed to any single nutrient.[46] A recent massive survey of the literature, moreover (with data on 68.5 million people) found that while overweight (body mass index [BMI] over 25) and obesity (BMI over 30) are—as is widely known—associated with cardiovascular and other diseases, the impact may be less important than is widely feared, and the authors of the study noted, "the rate of this increase has been attenuated owing to decreases in underlying rates of death from cardiovascular disease."[47] Indeed, life expectancies in the United States have continued to rise (as they have continued to rise in industrialized countries since 1840) by three months for every year lived, or six hours for every day lived, which is truly extraordinary.[48] Moreover, the so-called obesity paradox reveals that, under some circumstances, being overweight can apparently be healthful, and we should remember that as early as 1955 the WHO symposium noted that the Western diet had the dual effect of both stimulating and damaging our health.[49] We are still trying to understand these effects, and it is premature to confidently dictate our diet.

There is nonetheless currently a scientific consensus on health and diet, which has been summarized by the American Heart Association:

- Dietary saturated fat does increase the rate of cardiovascular disease significantly, (Keys got that right), but
- replacing it with refined carbohydrates and sugars does not reduce the rate of cardiovascular disease (Keys got that wrong, whereas his critics, including Yudkin, Mann, and Ahrens, got that right), but
- replacing saturated (animal) fat with unsaturated (vegetable) fat does reduce the rate of cardiovascular disease significantly (as effectively as treatment with statins) as long as those unsaturated fats are not trans fats.[50]

But this consensus has not yet deeply penetrated the public debate, and while most popular commentators now follow the Taubes anti-carbohydrate/anti-sugar story, the federal government and leading medical authorities still follow the traditional anti-fat story.[51] This divergence of opinion is both unnecessary and unhelpful.

A Nutritional Note

Reducing food to its constituent chemicals such as carbohydrate or fat is now increasingly criticized as "nutritionism," because doing so may mislead by ignoring the complex and generally unknown interactions between the different chemicals in food.[52] So, for example, the data may mean that meat is dangerous not because of its fat content but because of its protein or haem content (haem being the iron-containing chemical in meat that gives it a red color).[53] Equally, a Harvard group has identified that plant-based diets that are rich in sugars, starch, or refined carbohydrates may be unhealthful.[54]

The current American Heart Association's advice, therefore, seeks to avoid the nutritionist error by recommending a "Mediterranean" diet (rich in olive oil, vegetables, fruit, nuts, and legumes such as peas, beans, lentils, and chickpeas; moderate in fish, poultry, alcohol, and whole-grain cereals; and low in red meat, processed meat, and sweet foods such as cakes and jams). The similar DASH (Dietary Approaches to Stop Hypertension) diet is also recommended by the association.[55]

A Biochemical Note

Taubes suggested that, calorie for calorie, carbohydrates in the diet may promote obesity more than fat does because they stimulate the secretion of insulin, which in turn stimulates adipocytes (fat cells) to store fat: just as a girl entering puberty puts on weight around her hips and at her breasts because of the local actions on fat cells by certain female hormones (not because she has suddenly started to overeat), so someone who swaps fats for carbohydrates may start to put on weight because of the generalized effect on fat cells of insulin. Further, insulin drives down blood sugar levels, which in turn promotes the secretion of hunger hormones such as ghrelin, which therefore stimulates further eating.[56]

Conclusion

The central question remains: Why, in 1977, did the Senate Select Committee on Nutrition and Human Needs publish *Dietary Goals for the United States* when so many credible authorities, including Yudkin, Atkins, Ahrens, Mann, and the American Medical Association, had anticipated, at least in part, today's understanding of carbohydrates and other saturated fat substitutes as dangerous?

As Chapter 1 discussed, and as John Ioannidis, Brian Nosek, Paul Smaldino, and Richard McElreath have shown, one problem is that scientists are much less scientific than is popularly supposed. Aggravating the problem of poor science is that research operates a version of public-choice theory: in his 1965 *The Logic of Collective Action*, Mancur Olsen described how small interest groups can capture public policy, and publicly funded science is no exception.[57] Once armed with power over government-funded research grants, access to peer-reviewed journals, and appointments to university positions, Keys and his fellow elite researchers could enforce their paradigm on the whole field. Which was why the successful paradigm-shifter emerged not as a mainstream scientist but as a journalist, Taubes, who was spared the pressure to conform. And this public-choice aspect of science was aggravated by the Senate Select Committee's endorsement of the fat paradigm, which fueled it, literally, with public money.

The federal government's failings were further aggravated by lobbying. When McGovern's committee reported, the reaction from the meat, egg, and other food lobbyists was so vitriolic that the committee was forced to hold additional public hearings. Following those, a second edition of *Dietary Goals for the United States* was hurriedly released in late 1977, retracting some of its strongest earlier claims. For example, the committee added this sentence: "science [cannot] at this time ensure that an altered diet [will] provide improved protection from killer diseases such as heart disease."[58] As it happened, the meat and egg lobbyists were not wrong in all their objections, but they influenced the committee not because of their superior science, but because of their electoral power.

Because of the controversy, the reputation of the Select Committee suffered, and its mandate was allowed to lapse. The issuance of further federal dietary advice was then charged to the U.S. Department of Health and Human Services (HHS) and the U.S. Department of Agriculture (USDA), which have since jointly published, every five years, their *Dietary Guidelines for Americans* (the most recent was in 2015); these guidelines have, however, only reinforced the anti-fat, pro-carbohydrate message of the original 1977 *Dietary Goals*. This message was to be popularized in the USDA's Food Guide Pyramid (1992), MyPyramid (2005), and MyPlate (2011).

The USDA was so empowered specifically to help protect agricultural interests, which inevitably has distorted its advice. Marion Nestle, for example, professor of nutrition at New York University, reports that when she was hired to edit the 1988 *Surgeon General's Report on Nutrition and Health*: "My first day on the job, I was given the rules: no matter what the research indicated, the report could not recommend 'eat less meat' as a way to reduce the intake of saturated fat, nor could it suggest restrictions on intake of any other category of food."[59]

Indeed, Marion Nestle's book *Food Politics: How the Food Industry Influences Nutrition and Health*, was first published in 2002; it is now in its third edition, and it has been cited no fewer than 2,250 times, yet the food industry still influences the federal government's advice. Thus the *New York Times* reported on January 18, 2016, that the early drafts of the *2015–2020 Guidelines* confirmed that red and processed meats increase the risks of developing bowel and other cancers, but—following the lobbying of Congress by the National Cattlemen's Beef Association—the reference was removed from the final version.[60]

The problem of poor science is aggravated in food because of the vast weight of articles (containing highly selective information and highly selective statistics) that are published in distinguished journals by food companies themselves. A recent survey showed, for example, that reviews of the literature written by scientists with financial links to the sugar industry were *five times less likely* to conclude that sugar aggravated obesity or weight gain than did reviews by independent scientists.[61] Where a field of research such as food science can be dominated by producers' research, therefore, it can be doubly difficult to determine the various sources of bias. And the bias can be secret: we now know, for example, that Mark Hegsted, the Senate Select Committee's chief scientific adviser, was being secretly paid by the sugar producers to condemn fat and exonerate sugar and carbohydrates.[62]

Governments may be institutionally incapable of providing disinterested advice for at least four reasons. First, the scientists themselves may be divided, and by choosing one argument over another, the government may be making a mistake. Second, by abusing the precautionary principle, the government may be biasing its advice away from objectivity to risk-avoidance long before all the actual risks have been calculated. Third, because of public pressure, it may offer premature advice. And fourth, its advice will be distorted by lobbying.

Congress has lost patience with official dietary advice, so it commissioned the National Academy of Medicine to review "the entire process used" to generate the official guidelines, though unfortunately the review was limited only to methodology and did not include results.[63] But why should the federal government issue health advice at all? The general public, sadly, will give greater credence to federal government pronouncements than the science can bear, and it would surely be healthier to promote a free market in research ideas. There are a large number of scientific institutions qualified to give advice on food, and it would surely be healthier for the public to be exposed directly to their disagreements than for the federal government to proselytize an apparent consensus that is partial and selective.

The tragedy is that food science is, because of the power and money of its commercial sponsors, deeply flawed, yet government—by inserting its own biases—has only amplified that science's faults. Society does need a truly independent, truly high-powered entity to interrogate food science's output, but that will have to be sought among the ranks of people like Gary Taubes, who have made a career of probing the biases of science.

There is a tradition of politicians involving themselves in science. From William Jennings Bryan's attack on evolutionary theory in the Scopes Trial, to Al Gore and Donald Trump distorting modern climate science, politicians inevitably politicize science and almost always get it wrong. Senator McGovern's actions were only one of many examples that have damaged the credibility of both science and politics. It is an example of lessons we need to heed.

HEADS IN THE SAND

How Politics Created the Salt-Hypertension Myth

I n Chapter 2 we saw how science is often abused by powerful people with what they believe are good intentions. George McGovern, a heroic B-24 pilot in the European theater during World War II, no doubt had the best of intentions when his Senate committee provided harmful dietary advice about fat and carbohydrates. But he also had no problem cherry-picking incomplete science on salt and hypertension. Scientists who supported his position were called to testify in disproportionate numbers. In the House of Representatives, a young Rep. Albert Gore (D-TN) also condemned salt, also cherry-picking incomplete science to use against those who disagreed with him on global warming. As he did later with climate change, he claimed a scientific "consensus" that did not exist, except on his imbalanced witness panels.

In this book, most of the areas that we examine share many traits. One side will claim the moral high ground and will signal virtue while accusing its opponents of self-interest and corruption. "Settled science" is often invoked for some type of regulation, even if competing scientific research militates against action. And the incentive structure of modern science encourages a biased scientific canon in support of hypotheses that are in political favor. The corruption of modern science and its misuse by those in power extracts real costs, both personal and monetary.

Salt: The Spice of Life

We learned as schoolchildren that a person can live for weeks without food, but only a few days without water. This technically accurate trivia is also incomplete. Water is the most pressing of all humans' nutritive needs, but it is practically useless to us without the presence of another life-giving ingredient: salt.

Comprising two elements, sodium and chloride, humble table salt (NaCl) is required for a host of vital bodily functions, including tissue growth, bone formation, neural and nerve impulse transmission, muscle contraction, and reproduction. It also serves a more immediate physiological purpose. Salt, particularly its sodium ion, is the primary tool through which the body maintains fluid homeostasis. This balancing of our internal fluids depends on adequate supplies of both water and sodium.

Without sodium, the body cannot retain water and, as a result, cannot preserve the delicate balance of water and electrolytes in and around our cells. Nor can it maintain a high enough blood pressure for our kidneys to filter and supply clean blood to our vital organs. In other words, without enough sodium, everyday challenges like dehydration, fluctuating temperatures, and minor illness or injury suddenly become life-threatening events.

Because of salt's centrality to our survival, we have evolved a preference for salty-tasting foods. This penchant is known as "salt appetite," and it shifts as our internal levels of sodium rise and fall. Yet the amount of salt consumed by our species appears to remain relatively stable within a surprisingly narrow range. Despite differences in language, diet, culture, and socioeconomic status, the sodium intake of 95 percent of the world's human population is between 2,620 and 4,830 milligrams (mg) per day on average.[1] The average American falls almost exactly in the middle, eating about 3,400 mg of sodium per day.

While a difference of 2,200 mg might seem like a lot, it represents the equivalent of less than a teaspoon of salt. A majority of the population,

65 percent, consumes an even narrower range of 3,100 to 3,900 mg of sodium per day. Considering the ubiquity and cheapness of salt in most of the world, the narrowness of this range is impressive and coincides with the ranges observed in other mammals, such as omnivorous rats and herbivorous sheep (after accounting for animal size and metabolism).[2]

Not only is the worldwide sodium intake similar among humans, it also appears unchanged over time; Americans consume virtually the same amount of sodium today as in the 1950s (when salt consumption first began to be tracked).[3] This fact is notable because our intake of processed foods and calories increased during that time, along with the saltiness of processed foods, yet our sodium intake did not increase significantly. This consistency indicates that the amount of salt in our diets is not determined by food companies, as some argue. Rather, it appears that the level of salt we seek in our diets is set and tightly controlled by unconscious physiological mechanisms.

These mechanisms, though not thoroughly understood, have been observed in clinical settings. For example, in one double-blind clinical trial, patients on salt-restricted diets were given either a placebo or a salt tablet containing 2,300 mg of sodium. Clinicians noted that changes in dietary patterns, either avoiding additional salt or adding salt to foods, resulted in both groups' diets having virtually the same total sodium content by the end of the experiment.[4]

If salt is vital for our survival and the amount we crave is physiologically determined, how did salt come to be seen as a threat to health and the culprit behind all manner of maladies, including hypertension, heart disease, stroke, and obesity? The answer is that, while for most of our species' history salt was revered as a gift from the gods, 50 years of government-led anti-salt advocacy has poisoned not only our relationship with salt, but also the science of nutrition.[5]

Since before the U.S. government released its first *Dietary Guidelines for Americans* in 1980 and began telling Americans their intake of salt was high, a handful of strong-willed and well-connected experts had been waging a campaign to convince the world that salt causes hypertension and that salt restriction will save lives (the salt-hypertension hypothesis). This idea, while likely well-intentioned, stems from logical leaps made in the earliest days of research that were later proved incorrect. The idea that universal salt reduction leads to better health persists because of perverse incentives created by government interference in science and because many ignore or obscure the evidence that reducing salt is useless for most people and likely harmful to many.

This chapter details how these experts influenced government policy and how, through government involvement in the science of salt, they created the perverse incentives that led the research community to elevate an untested hypothesis to the level of dogmatic truth. The chapter is separated into four sections. The first section covers the early period of research and the origins of the salt-hypertension hypothesis. The second section details how experts forced this tenuous evidence into public policy and how they fueled the propaganda that led to the widespread belief in the salt-hypertension hypothesis. The third section describes the ongoing campaign to preserve "consensus," which modern research increasingly threatens. The fourth section discusses the effects the war on salt has had on public health and what should be done to fix it.

Origins of the Salt-Hypertension Hypothesis

The human body needs sodium in order to elevate blood pressure. Thus, to a certain extent, the theory that *too much* sodium leads to blood pressure that is too high makes logical sense. But the sophisticated processes underlying this mechanism also explain why the theory is largely incorrect.

Most of us are well aware of the long-term risks posed by very high blood pressure, but the much more immediate threat to our survival is blood pressure that drops *too low*. When the body suffers a significant loss of blood pressure, the kidneys cannot clean waste from the blood fast enough to supply our vital organs with fresh, oxygenated blood to keep them, and us, alive. And because the bulk of our blood is composed of water, the greatest threat to blood pressure stability is dehydration.

To understand what happens inside the body during periods of dehydration, imagine that your blood vessels are like a garden hose and your blood like the water flowing through it. Because our blood is made up largely of water, dehydration causes blood volume to decrease. As slowly turning a spigot off causes a hose to slacken and the water flowing through it to come out more slowly, a decrease in blood volume reduces the rate or pressure of our blood flow. When the body senses such a drop, it triggers two main responses. First, it releases hormones that make our blood vessels constrict or tighten. Just as putting a thumb over a hose opening causes water to flow out more quickly, vessel constriction reduces the space through which blood can flow, causing its pressure and speed to increase. However, this mechanism will not save us for long if the body continues to lose water and blood volume. So, in order to preserve the volume it has, the body also triggers water retention. And the first way the body retains water is by holding on to its sodium.

The need to conserve sodium in order to preserve water stems from the physiological axiom that *water follows salt*. By signaling the kidneys to reabsorb sodium, instead of releasing it into urine, it allows the kidneys to also reabsorb water rather than excrete it. By retaining water, the body can prevent further decreases in blood volume, and thus blood pressure. This process also triggers the sensation of thirst, prompting us to replenish lost water and restore blood volume to its normal level. Like opening the spigot on a faucet, in the human body, increased water intake combined with water retention increases the pressure and speed of blood flow.

It is this mechanism for preserving fluid homeostasis—the retention of sodium and water to elevate decreased blood pressure—that led some early researchers to the salt-hypertension hypothesis: the idea that excessive dietary salt causes chronically elevated blood pressure. But the purpose of this system is not simply to *increase* blood pressure; it is to preserve *balance*. Not until later in the history of scientific inquiry would researchers discover that, as with sudden drops in pressure, sudden increases in pressure trigger a counter-response and the release of hormones that allow blood vessels to relax; increases in pressure also stimulate the kidneys to excrete sodium and water in order to lower blood pressure and achieve the body's primary goal: homeostasis.[6]

Unfortunately, an understanding of the body's counter-responses to high blood pressure and sodium levels didn't come until the 1980s. And it wasn't until later that scientists began investigating what the "optimal" level of salt intake in humans might be and whether sodium deficiency might be worse than salt "excesses." By then, too many careers and egos were invested in the salt-hypertension hypothesis to admit that salt restriction has no benefit for most people and has the potential for significant harm for many.

So we have, in the salt debacle, public health held hostage by the incentive structure for modern science. Researchers have no incentive to disprove the hypotheses that have served as the basis for enough research funding to gain tenure. Doing so would obviously impair their future advancement.

Salt and Disease

For at least 2,000 years, physicians suspected that salt played some role in health. But it wasn't until the 20th century that clinical researchers began to investigate a possible connection between salt and blood pressure. They began, where most medicine does, with the ill.

In the early 1900s, after observing that many patients with heart and kidney ailments also often experienced edema (swelling of body parts due to water retention), French physicians Ambard and Beaujard conducted one of the earliest experiments in salt restriction. By putting their hypertensive patients on low-salt diets, the pair found that this not only reduced some of the swelling, but also seemed to decrease blood pressure. Though admittedly "not striking" nor consistent, their results piqued the first wave of interest in studying the salt-hypertension connection.[7]

In a pattern that will be repeated throughout the history of salt research, those who sought to replicate the work of Ambard and Beaujard obtained mixed results, and many of them concluded that the treatment was beneficial to only a minority of hypertensives.

The hypothesis then largely fell out of favor until the 1920s, when Frederick Allen, "one of the most outspoken advocates of salt restriction," began testing his hypothesis that hypertension was caused by a

FIGURE 3.1
Correlation of average daily salt (NaCl) intakes with prevalence of hypertension in different geographic areas and among different races

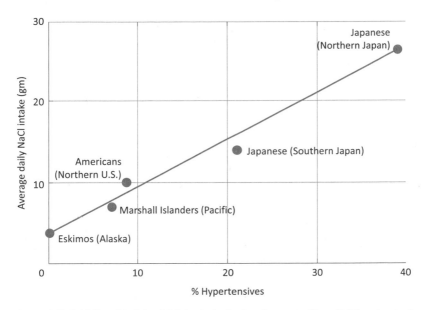

Source: L. K. Dahl, "Possible Role of Salt Intake in the Development of Essential Hypertension," *International Journal of Epidemiology* 34, no. 5 (October 2005): 971, fig. 1.

When they graphed the data for their five populations, it produced a striking linear trend: blood pressure rose in line with salt intake. This led the authors to conclude that "with higher average salt consumption there is higher prevalence of hypertension."[19] The picture they painted practically begged readers to jump to the conclusion that salt *caused* hypertension, but what it actually showed was merely that the two variables *correlated* with each other.[20]

As Dahl himself noted, just because two variables appear to rise or fall in line with each other (in this instance, salt and hypertension), does not necessarily mean that one *causes* the other. Put more succinctly, this can be phrased as the statistical maxim: correlation does not prove causation.

A correlation, like the one observed by Dahl and Love, *may* indicate some relationship between the two variables; it does not prove it.

Often such relationships represent something known as a spurious correlation—in which the link between the two is entirely coincidental. For example, one could rightly say that the divorce rate in Maine correlates almost one-to-one with the per capita consumption of margarine in the United States. That is, plotted on a graph the rate of divorce in Maine would rise or fall proportionally to the level of margarine consumed. This correlation, while accurate, does not mean that Americans who eat more margarine *cause* more Maine couples to divorce. Nor does it mean that a sudden rise in divorces in Maine *causes* Americans to eat more margarine. It is possible these two variables are related, but merely demonstrating a correlation does not show how or if they influence each other, only that they changed at the same time.[21]

Because it is unlikely that Maine divorces have anything to do with the nation's margarine consumption, this correlation is obviously spurious. It is more likely that other factors, not considered, are what influence rises and falls in margarine consumption or divorce, or both. These other factors are known as "confounders" or confounding factors. One way epidemiologists attempt to more accurately determine correlations that are not spurious, and that demonstrate some sort of causative relationship, is by accounting for and factoring out potential confounders. Doing this in health research is particularly difficult because the factors that influence both health and dietary intake are virtually unlimited and are often overlapping. For example, having a lower income makes one both more likely to eat a poor diet and less likely to have access to healthcare. Thus, even if diet is unrelated to health, those in lower socioeconomic strata may have worse health outcomes. Even if diet is independent of these health outcomes (though it likely isn't entirely so), failing to account for the confounder of access to healthcare may bias the correlation.

When it comes to hypertension, a major confounding factor is age. The disease usually does not develop until middle age or later, so a group of 50-year-olds in Finland will show higher rates of hypertension than a group of 20-year-olds in Sweden, regardless of salt intake. When comparing primitive cultures to westernized cultures, as Dahl did, age is a confounder because industrialized populations tend both to consume more food and salt *and* to live longer. Thus industrialized populations will necessarily appear to have a higher incidence of hypertension and salt intake, even if these factors are independent of each other.

Another well-documented contributing factor in the development of hypertension is body weight. Dahl himself noted in 1954 that hypertensives were twice as likely to be overweight as their nonhypertensive (normotensive) counterparts.[22] Furthermore, since overweight individuals tend to eat more food overall, they are also likely to consume more salt than those on lower-calorie diets. Thus, if the primitive populations Dahl studied were of lower weight—a likely condition considering the scarcity tribal societies often endure—the industrialized populations would necessarily have both higher blood pressure and higher salt intake.

In order to confidently assert that two variables that correlate are likely related, researchers need to account for and eliminate these other potential confounding factors. Unfortunately, although Dahl and Love did report the average age of their populations, it appears they did not control for age and took no other possible confounders into account. This lack of detail, though regrettable, was not uncommon for the era and remains a problem for researchers working with limited resources. Furthermore, many of the other risk factors for hypertension we know about now were unknown or unproven at the time.

Apart from lifestyle and dietary influences on hypertension, another factor that might explain variation in blood pressure among the groups is genetics. Some cultures might have an inherited predisposition to developing the disease, no matter how much or how little salt they eat. Dahl acknowledged this as a possible factor, writing in 1960 that "there is considerable evidence which suggests that hereditary factors operate in human hypertension."[23]

Excited by what Dahl had found, other researchers sought to validate his results and account for the many possible confounding factors not included in Dahl's study. While science on genetics was limited at the time, there was a method to somewhat limit the chances that confounding factors might bias the results: by studying blood pressure and salt intake *within* a given population. If salt truly were driving the blood pressure rates, then one would expect to find the same correlation among people living in a given area (known as "intrapopulation" studies). Since the sample populations in these intrapopulation studies would share a similar distribution of culture and genetics, if higher salt intake here still correlated with higher blood pressure, one could say with more certainty that salt was indeed a driving factor behind hypertension. This, however, is not what researchers found when they conducted these intrapopulation studies.

SALT AND HYPERTENSION WITHIN GROUPS

In the decades following Dahl's five-population study, there were dozens of attempts to replicate his findings within a given population, but none found a clear link between salt and blood pressure.[24]

Beginning in 1937, studies that examined the diets of normotensive individuals found no significant difference between the amount of salt they ate and the amount consumed by those with hypertension. This includes a study Dahl himself conducted, in which he looked at diet and blood pressure for workers at his laboratory.[25]

One of the longest running of these intrapopulation assessments is the so-called Framingham Study. Started in 1948, it has continuously tracked the behaviors and health of thousands of participants living in Framingham, Massachusetts, with the goal of identifying potential causes of chronic illnesses like heart disease. In its 1967 report, the study concluded that there was no appreciable connection between blood pressure and 24-hour urine sodium excretion, a method of estimating sodium intake.[26] That lack of connection was confirmed again in 2017 when the Framingham Study—now on its third generation of subjects—found that of the more than 2,500 individuals followed over 16 years, lower sodium intake was not associated with lower blood pressure. In fact, they found the opposite: those who consumed lower levels of sodium, less than 2,500 mg a day, tended to have *higher* blood pressure.[27]

Wisely, as the body of evidence continuously failed to show a direct link between blood pressure and salt, Dahl updated his hypothesis. Even while most individuals on a high-salt diet will never become hypertensive, it doesn't prove salt isn't involved in the disease, he reasoned, "rather that not *only* is salt involved." This, he asserted, is similar

lobbying government to institute programs and policies aimed at protecting consumers from corporate interests. It was within this milieu that Dahl began promoting the idea that the high salt content of commercially prepared foods was a cause of hypertension. His campaign began, as most modern public health campaigns do, with children.

Dahl's Rhetoric

While Dahl's experiments did not conclusively show a link between salt and hypertension, they did provide him with one significant finding: the younger the rats were when he initiated salt-loading, the more likely it seemed the animals' blood pressure would rise. Even if rats were later put on a normal salt diet, Dahl found that weanlings fed high-salt diets were more likely to develop hypertension. This was the basis for Dahl's argument that the "high" salt content of commercially prepared baby foods was not only unnecessary, but potentially deadly.[37]

Human breast milk, Dahl argued, is a naturally low-sodium food, and infants fed breast-milk-only diets were perfectly healthy without any signs of sodium deficiency. Thus, he reasoned, there was no need to add salt to commercial baby foods. Although he conceded that most children likely would not be harmed by the salt in baby foods, he argued that for salt-sensitive babies, like his rats, the added salt might put them at risk for hypertension later in life, and therefore it should be eliminated from all processed baby foods.

In what would become a permanent feature of the anti-salt movement, Dahl argued that reducing salt intake in entire populations would not only benefit the small portion of those at risk for hypertension, but would also not harm the rest. "One might say that is only the hypertension-prone child that we need worry about," he wrote in 1968, but there is no way to tell which children might be salt-sensitive, "therefore, salt should be decreased for all."[38]

But not all in the research community agreed with this assumption or the tactics Dahl employed to push his message to the public. In one study, Dahl and his colleagues reported that they were able to induce hypertension in some of their salt-sensitive rats merely by feeding them commercially available baby foods. Human infants, of course, are much larger than rats and therefore can handle much higher sodium levels.

Forced to address the public fear stoked by Dahl's study, Charles U. Lowe, chairman of the Committee on Nutrition at the Academy of Pediatrics, wrote that it "smacks of the scare techniques which might be used in advertising."[39] Dr. Henry Schroeder, professor of physiology at Dartmouth Medical School, noted that even with Dahl's work there was "no evidence one way or the other" about the potential harm of salt in baby foods, or for the benefits of salt restriction for hypertension. Schroeder also raised the point that removing salt from baby foods, far from being harmless, might actually put children at greater risk under certain conditions. The ultra-cautious doctor might advise low-sodium diets for children with a family history of hypertension, Schroeder wrote, but "the average pediatrician might be grateful for the extra margin of safety against dehydration should his patient contract severe diarrhea."[40]

These concerns, however, were largely ignored when Congress waded into the issue in 1969. In fact, Dahl was invited to testify when the Senate Select Committee on Nutrition and Human Needs held hearings on the issue; this is the same committee, chaired by the same senator, George McGovern that would be responsible for the misleading dietary guidelines on fat and carbohydrates. Dahl warned the committee members that their failure to act on salt could cost lives. "It took many years and numerous tragedies to develop an awareness of the baneful effects . . . from exposure to apparently benign environmental influences," like pesticides, cigarettes, and smog, Dahl testified. He claimed that the results of his and other studies showed ample evidence of a connection between salt and hypertension and that "this is a time for caution rather than plowing ahead blindly as we have been doing for a quarter of a century."

In another tactic that would become a favorite of public health advocates, Dahl insinuated throughout his testimony that large corporations were harming public health in service of profits and that it was government's job to intervene. Parents, he noted, often taste-test their children's foods, and they would reject unsalted foods as unpalatable. "The salt added by food processes apparently is added primarily to enhance sales appeal to mothers and is not based on estimates of infants' needs," Dahl testified. This specter of corporate greed would prove highly effective and popular, not only for those advocating government action on salt, but also as a means of discrediting any research that questioned the wisdom of universal salt reduction.[41]

Whether intentionally or not, by targeting children, Dahl had incited widespread concern over the level of salt in the American diet. While some researchers publicly denounced Dahl's claims, they could do little to temper the public fear stoked by the government inquiry. As a result, the issue of salt and health would be put through a battery of investigations by both government agencies and nongovernmental health organizations. And these inquiries, though they produced mixed results, only served to further increase public skepticism about the healthfulness of commercial foods and increase the pressure for regulatory action.

Salt Debate Enters the Government Arena

Throughout the 1960s, the evidence and views of the research community remained stubbornly inconclusive on the issue of salt. Yet many key figures had been swayed by Dahl's research. One of these was Dr. Jean Mayer, scientific adviser to President Richard Nixon. In 1969, Nixon chose Mayer to lead the White House White House Conference on Food, Nutrition and Health, a symposium of the nation's top nutrition experts convened to review the existing evidence and reach a consensus about the top health questions of the day.[42] Dahl was among those invited by Mayer to participate in the symposium, as was anti-fat proponent Ancel Keys (a key influence on dietary guidelines, as discussed in Chapter 2). Though the conference's focus was on hunger and malnutrition, it frequently turned to the problem of "overnutrition." It also served as an arena in which the increasingly acrimonious salt debate would play out in full view of the public.

Two panels considered the issue of dietary salt. One, on which Dahl served as a member, concluded that reducing the salt content of infant foods was "desirable." The other, chaired by Lowe, determined that there was "no scientific basis" for that recommendation. Breast-milk-only diets did provide children with adequate nutrition, Lowe's panel noted. However, added salt may serve "a major physiological need" as children transition from breast milk to solid food diets. This is because breast milk, while low in sodium, is also low in potassium. Infant foods, on the other hand, which often consist of fruits and vegetables, are high in potassium. This is important because sodium is necessary to clear potassium from the body. Lowe's panel worried that eliminating salt could lead to a dangerous buildup of potassium in children. "This and other potential hazards from low sodium and high potassium diets must be weighted against the possibility of predisposing the infant to hypertension in later life," Lowe's panel wrote, and concluded that they had found "no data indicating that the current intake of sodium by the human infant has a hypertensive effect."[43]

But Dahl did not recognize the potential benefit of consuming anything but the absolute minimum of salt needed for survival, writing that "there is no evidence—physical, physiological, biochemical, or psychological –to support such a theory."[44] However, previous salt-restriction experiments had provided ample evidence that consuming such a low-salt diet in fact had numerous potential health hazards.

As previously discussed, Herbert Chasis in 1950 had found that nearly all the healthy people he put on low-salt diets experienced significant declines in kidney function. Even as far back as 1935, researcher Robert McCance of King's College London, who had put himself and several research assistants on a low-salt diet, reported concerning side effects that included severe cramps, lethargy, anorexia, and a buildup of urea in their blood.[45]

Dahl's willingness to disregard any evidence indicating that salt restriction was neither wholly beneficial nor entirely harmless would become the modus operandi for the anti-salt movement. In fact, by the 1970s those experts already committed to the salt-hypertension hypothesis would not only ignore, but systematically work to discredit any study that indicated reducing salt in the general population could increase health risks. Unfortunately, government relied almost exclusively on these researchers as policymakers debated what dietary recommendations to provide to the entire nation.

Government Enters the Salt Debate

In the years after the White House conference, the debate on salt had developed into an unending volleyball match between the opposing sides. Interest in the topic had never been higher, triggering a landslide of studies seeking to determine whether salt caused hypertension and whether salt reduction benefited certain individuals. Yet for every study published that touted evidence of salt restriction's benefits, there was another that refuted the idea. Rather than illuminating the role of salt in health, each passing year seemed to make the issue even murkier. This back-and-forth, while common in scientific debates, began around this time to take on a few distinctive features that would later lead journalist Gary Taubes to describe it as "one of the longest running, most vitriolic, and surreal disputes in all of medicine."[46]

For one, the tenor of the discussion became decidedly biased. Instead of considering the merits of the growing body of research, individuals in the opposing and increasingly calcified camps began simply to ignore or dismiss evidence unsupportive of their side. Worse, some participants engaged in ad hominem attacks, openly criticizing the character and motives of researchers on the opposing side. This was particularly true of those in the anti-salt camp, who defended universal salt reduction with an almost religious fervor.[47]

In 1979, Simpson commented on these disturbing undertones in the salt debate, writing, "We seem . . . to have got into a situation where the most slender piece of evidence in favour of [the salt-hypertension hypothesis] is welcomed as further proof of the link, while failure to find such evidence is explained away by one means or another."[48] Seeming to illustrate his point, that year cardiologist Jeremiah Stamler and his colleagues at Northwestern University Medical School began publishing the results of their trials examining salt intake and blood pressure in Chicago-area schoolchildren. While the first of these tests found a "clear-cut" link, two follow-up attempts failed to reproduce these results. Despite their inability to reproduce their initial results, the team described their trials as "not wholly negative," and provided numerous explanations for why they were unable to reproduce their initial results, including that the length of the study was too short, genetic factors might have obscured the link, or their instruments might not have been sufficiently sensitive to detect differences.[49]

But Stamler's trial was just another in a long line of intrapopulation studies that had failed to show a link between salt and blood pressure within populations, a trend that would continue. In 1981, for example, a study of 120 randomly selected 49-year-old men in Gothenburg,

Sweden, showed no connection between sodium intake and blood pressure, leading the authors to conclude that it did "not lend support to the hypothesis that habitual salt intake might be of major importance for the blood pressure level in mild to moderate essential hypertension."[50]

Those in the anti-salt camp dismissed these results. "This finding is not unexpected since the range of intra-individual salt intake in most Western countries is small," Graham MacGregor, a researcher at Charing Cross Hospital in London and a leader of the salt restriction movement, wrote in 1985. A single 24-hour sample of urine is not adequate to estimate a person's long-term habitual salt intake, he argued, since it may it may vary from day to day. He also asserted that even collecting a week's worth of urine, as Stamler's study had, would not reflect lifetime intake habits, so that, as Dahl argued, a child who consumed a very high-salt diet might go on to consume less and still develop hypertension as an adult. These failures, he asserted, do not refute the preponderance of evidence, which suggests that low-salt diets can reduce blood pressure.[51] This logic was emblematic of what Simpson called the "resilience and virtual indestructibility of the salt-hypertension hypothesis." Negative results, he noted "can always be explained away."[52]

In an award-winning 1998 article for *Science*, Gary Taubes theorized that the reason the salt debate had become "one of the longest running, most vitriolic, and surreal disputes in all of medicine," stemmed from the fact that epidemiological tools used to gauge salt's influence on health were "incapable of distinguishing a small benefit from no benefit or even from a small adverse effect."[53] In other words, while many studies could find a *statistically* significant connection between salt and blood pressure, they could not find a *clinically* meaningful relationship. That is, even if studies found blood pressure rises of one or two points as salt is increased by 1,000 or 2,000 mg a day, they did not provide clear evidence of what this means for health, leaving researchers free to claim a given study did or did not support their side of the debate. As Yale University School of Medicine epidemiologist Alvan R. Feinstein wrote in 1985, the war was being fought over "distinctions that seem basically insignificant." The link between salt and blood pressure that researchers were sometimes able to find was "too small to receive serious consideration as issues of scientific importance." They only became significant, Feinstein argued, because clinicians "engaged in a type of intellectual lobotomy that equates statistical significance with biological, physiological, or quantitative importance."[54]

As the literature on salt continued to pile up, both sides of the debate had claim to what Stamler referred to as a "totality of data"—a body of research so large that it seems to show unambiguous proof for one's position. Those in the anti-salt camp could point to hundreds of studies that seemed to indicate high salt intake caused hypertension. But the "totality" of evidence supporting universal salt reduction existed only for those ignorant of the equally large body of data contradicting it. Unfortunately, the members of government who set out to answer the salt debate once and for all fell into this category. And it was this government interference that would skew research on salt from then on.

In the scientific process, the proper referee for competing hypotheses is time. Decades of research evaluated by members of the scientific community will eventually lean toward one explanation or another. As this happens, researchers shift energy and financial resources toward investigating those hypotheses with the most promise. However, when government, which is the primary funder of science, and often the final word, acts as a sort of referee, it creates a host of perverse incentives that can influence which hypotheses researchers choose to favor.

Government as Science Referee

Unlike researchers, politicians and bureaucrats are usually uninterested in hearing about the limitations of the existing evidence. "They don't want the answer after we finish a study in five years. They want it now. No equivocation," Bill Harlan, director of the Office of Disease Prevention at the NIH, commented in 1998.[55] Thus, as Congress and government health agencies set out to create nationwide recommendations on salt, the experts they chose to listen to were those willing to unambiguously declare that drastic reductions in salt would improve Americans' health, no matter how shaky the evidence underlying their assertions.

The attention and conflicting conclusions that resulted from the White House conference in 1969 compelled many health organizations to weigh in on the issue. In 1971, the Food and Drug Administration (FDA) asked the Food and Nutrition Board (FNB) of the National Academy of Sciences to investigate the existing science and make a recommendation about salt in infant foods. While the FNB found "no valid scientific evidence available in support of the contention that addition of salt to infant foods contributes to development of hypertension or other disease states in adult life," the organization felt comfortable enough to recommend a modest 25 percent reduction, since the amount of salt in commercial foods was significantly greater than the estimated minimum daily requirements.[56]

This report, and growing public concern, spurred the FDA to call a meeting with the three largest producers of baby foods to convince them to voluntarily reduce sodium in their products. Aware of the public attention and looming threat of mandatory regulations, Gerber, Heinz, and Beech-Nut Nutrition Company quickly acquiesced to the government's request and quietly reformulated their infant foods to include less salt. Then in 1977, Beech-Nut, the smallest of the big three, began promoting its low-salt and no-salt-added foods as a means of bolstering faltering sales. The advertising war this triggered, far from convincing consumers of the safety of infant foods, instead had the opposite effect.

According to industry analysts, mothers' concerns over salt in baby food reached their apex in the late 1970s as the companies advertised their low and no-salt-added foods.[57] Their efforts to allay consumers' fears, which included funding investigations into Dahl's hypothesis, merely nurtured greater skepticism about the food industry and the healthfulness of the nation's food supply. It also reinforced the case made by public interest groups like CSPI, which had begun calling on government to restrict salt in the food supply.

The Senate Select Committee on Nutrition and Human Needs, chaired by George McGovern (D-SD) held hearings on the health of the American people in the 1960s. Initially focused on hunger and malnutrition, by the 1970s it had turned its attention to the role the "rich" American diet played in the development of chronic diseases.

McGovern took a particular interest in the causes of heart disease after receiving disturbingly high cholesterol readings from his doctor. By 1972, he had fallen under the influence of diet guru Nathan Pritikin, whose austere diet—low in fat and free of sugar, meat, added salt, or processed foods—promised to save clients from the ravages of age-related ailments. By following Pritikin's diet and exercise regimen (which included running three miles a day), McGovern was able to reduce his cholesterol impressively, from 350 to 170 milligrams per deciliter. He became a lifelong adherent of Pritikin's diet and a fervent believer that the Western diet was deadly.[58] It also made him a receptive audience for Dahl, who testified before McGovern's committee that the nation's high rates of hypertension were a result of an "unrestrained addition of salt." Though Dahl admitted to the committee that evidence proving a connection between salt and hypertension was "not unequivocal," he argued there was enough, including his studies on rats, to put "the burden of proof [on] the industry to prove it is not harmful."[59]

Fat was by far the more contentious nutrient considered by McGovern's committee, but the three days of testimony served to highlight the division within the scientific community on the evidence implicating salt in the development of hypertension and the benefits of salt restriction for the general population.

Some witnesses, like E. Cowles Andrus, a cardiologist at Johns Hopkins Hospital and former president of the American Heart Association, told the committee that researchers were "a long way from understanding" hypertension and that it was "not alone due to salt," but Congress was far more interested in the clear-cut pronouncements provided by the testimony of Dahl, Pritikin, Stamler, and others who told them reducing salt in the American diet would save lives.[60]

The McGovern Committee's report, "Dietary Goals for the United States" (the same report discussed in Chapter 2 that recommended the low-fat, high-carbohydrate diet), advised adults, among other things, to limit salt intake to under less than three grams per day (1,164 mg of sodium)—a 50–85 percent reduction based on estimates of average intake at the time.[61] It sparked an immediate and fierce backlash.

Predictably, food companies viewed the government's recommendation to cut fat, cholesterol, and salt from the diet as a threat to their businesses. But they were not alone; the report enraged many in the research and medical communities. "All hell broke loose. . . . Practically nobody was in favor of [McGovern's recommendations]," noted Mark Hegsted, head of the U.S. Department of Agriculture (USDA) Center for Human Nutrition and main adviser for the McGovern Committee Report.[62] Those experts who had previously declined to enter the political arena began voicing their opinion on what they saw as improper government interference in health issues. Many of the harshest criticisms centered on the evidence, or lack thereof, underlying the recommendations.

In a scathing analysis of the dietary goals, University of Wisconsin biochemist Alfred Harper wrote that they "appeal to those who accepted pseudoscientific reasoning."[63] The American Medical Association (AMA) concluded they were based on evidence that was "not conclusive" and had "potential for harmful effects."[64] And Philip Handler, president of the National Academy of Sciences, summed up the recommendations simply as "nonsense."[65]

Harper's 1978 analysis, published in the *American Journal of Clinical Nutrition*, condemned the goals for being scientifically unsound and noted that "it is a political and moralistic document" where the conclusions were "preconceived and then evidence was marshalled to support them." Harper was concerned about the potential harm the recommendations might have on public health; he also feared that they could undermine faith in nutritional advice altogether. Like the "food faddists and . . . necromancers of old," the goals, Harper wrote, seemed to promise a panacea for the disease of old age, providing no information about the likelihood of their success or potential risks. He feared that recommendations like these, based on pseudoscience, would jeopardize the public's trust in both government and health authorities and might hamper future education efforts on nutritional issues.[66]

Although critics of the goals pointed specifically to the lack of evidence on fat, many noted that the science underlying salt restriction was similarly unsound. Of great concern to many was the fact that the committee had relied primarily on trials conducted on individuals with hypertension and other adverse health conditions. This, some argued, provided no evidence of how the substantial salt reductions recommended by the goals might affect the general public. The benefits for the normotensive population appeared minimal at best, and the risks were completely unknown. Even those who testified in support of salt reduction recognized this danger.

Robert N. Butler, the director of the National Institute on Aging at the NIH, testified before the committee that salt restriction might help prevent the development of hypertension, but that "evidence is not available" for its benefit in the general population.[67] In a more colorful description, Harper compared advice to the public to consume salt levels shown to benefit hypertensives as akin to "recommending that protein intake of the U.S. population should be only 35 grams a day," a level that "is recommended for patients with renal failure."[68]

Among those who supported salt reduction, some, such as the Food and Nutrition Board of the National Academy of Sciences, took issue with the extraordinarily low level selected by the committee. While the FNB had felt comfortable recommending a 25 percent reduction in salt in baby foods, its chairman, Gilbert A. Leveille, later noted that the three gram limit was not supported by the evidence. The idea that this enormous reduction in salt would reduce the nation's blood pressure was based on "a modicum of tenuous information," Leveille testified. He also noted that "virtually all professionals examining the dietary goals of the Select Committee are in agreement that the recommended level of salt intake of three grams per day is excessively low."[69]

"I think [nutritionists] felt that a Senate committee had no business getting involved in recommendations that ought to be made by the scientific community," Hegsted said, but insisted that there were "no identifiable nutritional risks associated with shifting our diet," as recommended by the goals.[70]

The intensity of the outrage ultimately compelled Congress to hold a series of follow-up hearings. But instead of generating consensus on government nutritional guidelines as the committee had hoped, the additional testimony only served to highlight how deeply divided the scientific community was on the issue of salt.

Three Camps Emerge

The original witnesses called by McGovern's committee were almost exclusively in favor of the government's recommendations, creating an apparent "consensus." But the response the dietary goals generated made it clear that the scientific community was in agreement on almost none of the issues considered by the committee. It appeared scientists were divided into three distinct camps:

1. those who thought sweeping dietary changes would be beneficial;
2. those who believed such changes were unnecessary or harmful; and
3. those who simply believed it would be unwise to offer dietary prescriptions based on the existing scientific data.[71]

Those taking a more moderate position on the goals were, necessarily, far less vocal than those in support and thus less convincing to the members of the Senate committee as it held follow-up hearings. Robert Olson, chairman of the Biochemistry Department and professor of medicine at St. Louis University School of Medicine, begged the members to wait for better evidence. Like Harper, Olson feared that offering advice to the American public that was later deemed incorrect would be "worse than none at all." But his pleas fell on deaf ears. McGovern responded to Olson by declaring that "senators don't have the luxury that a research scientist does of waiting until every last shred of evidence is in. When we get the kind of overwhelming consensus that has developed before this committee . . . we have an obligation to share that with the American people."[72]

But as George M. Briggs, a University of California, Berkeley, nutritionist, pointed out, there was no consensus. The illusion of agreement arose from the committee having selected a list of witnesses who were overwhelmingly in favor of their recommendations. The goals, he argued, were based on "very limited formation from the testimony of a rather small number of selected persons, not all of whom are trained or experienced in giving dietary advice to the American public."[73] And even if consensus did exist, Dr. Edward Ahrens from the Rockefeller Institute of Medical Research argued that was not reason enough to make sweeping pronouncements on the American diet. The issue should be based on evidence and not "settled by anything that smacks of a Gallup poll," he testified.[74]

Rather than heeding the warnings of these cooler heads, the McGovern committee was charmed by those experts telling them that it was both their right and their responsibility to make population-wide recommendations. Stamler told the committee members that while there was no "precise" evidence that Americans were eating more salt, there was "an impression of rising intake." Furthermore, he noted that even if salt was not the sole cause of hypertension there was mounting evidence that high intake "helps set the stage" for individuals predisposed to the disease. Like Dahl before him, Stamler dismissed the argument that salt restriction ought to be a recommendation that only doctors make to their patients with hypertension. The disease, he argued, should be viewed as an infectious disease, like cholera. In such a scenario, treatment cannot be left to the "individual doctor-patient relationship," Stamler warned.[75]

The second edition of the McGovern Committee Report, released in December 1977, made only modest changes to the original recommendations. Among these, it raised the daily salt limit from three to five grams a day (1,164 mg to 1,940 mg of sodium). In a letter to the industry, McGovern also clarified that this limit was not meant for *total* salt intake, but rather for salt added to raw and commercially prepared foods.[76] If that is what the committee truly intended, then the total salt limit it actually recommended (accounting for the salt naturally present in raw foods) was nearer to *eight* grams a day, or 3,100 mg of sodium. This would not have been far from what most Americans ate, which estimates at the time put at between 10 and 12 grams (3,880–4,650 mg of sodium).[77] Whatever the committee's intent, however, the rest of the public health community took the new goals to mean that limiting total salt to under five grams a day would reduce hypertension risk and set about the task of enshrining the theory as dogmatic truth in the public's mind.

Consensus- and Career-Building

Following the publication of the second edition of the McGovern Committee Report, health organizations began releasing their own versions of nutritional guidelines. Although Robert Levy, as director of the National Heart, Lung, and Blood Institute (NHLBI), had testified in 1977 that more research was needed to demonstrate "the efficacy of salt lowering in the American free-living population," NHLBI in 1979 officially backed McGovern's five-gram limit.[78]

Similarly, while Harper, chairman of the Food and Nutrition Board at the National Academy of Sciences, had condemned the dietary goals as "pseudoscience," in 1978, only two years later, the FNB issued a report echoing McGovern's recommendations and advising adults to limit salt intake to no more than three to eight grams a day.[79] Likewise, though the Surgeon General's Office noted in 1979 that salt's role in hypertension was "not yet completely understood," a year later it felt comfortable enough to endorse the five-gram limit.[80]

It was not that the science had suddenly become clearer in the two years following McGovern's hearings. Rather, it was the mere fact that the government had made definitive pronouncements on diet and health that compelled health organizations to follow suit. They viewed health advice as *their* purview, and their failure to provide the American public with quantitative guidelines as McGovern had done would be seen as an abdication of their authority. They had to take a stance on something and, as one participant in these deliberations later commented, "salt seemed to be the easier target."[81] Thus, when these health organizations felt the need to reassert their authority on nutrition, salt recommendations were where many started. Their backing gave the impression that a consensus on salt had finally been reached. And though it was the result of a glorified turf war, it lent support to those advocates pushing government to regulate salt in food.

Al Gore and Sodium Labeling Hearings

Over the years, the FDA had debated the idea of requiring food manufacturers to list sodium content on labels—a suggestion Dahl's committee made in 1969. It wasn't until the dietary goals were established and the subsequent consensus of health organizations that the agency decided to put its plan into action. Although the election of Ronald Reagan in 1981 and his commitment to "regulatory voluntarism" seemed to pose a hurdle to such a new mandate, it actually provided an incentive for Democratic lawmakers to take up the issue. Indeed, championing the cause of sodium labeling would turn out to be a golden opportunity for Reagan opponents to demonstrate that, unlike the administration, they cared enough about public health to stand up to industry.

One such lawmaker was Rep. Al Gore, who at the time served as chairman of the House Science Subcommittee on Oversight and had a longstanding interest in nutrition. His committee had no legislative authority, but it served as a platform from which members could promote pet issues—and their own careers. As it happened, the issue of salt in the diet was also a topic of great interest to the man in charge of setting the committee's agenda: Thomas Grumbly, staff director of Gore's committee.

Previously, Grumbly had worked on nutrition issues at both the USDA and the FDA. At the FDA, Grumbly worked with CSPI when the group petitioned the agency for sodium labeling. It is possible that this relationship convinced Grumbly of the need to bring sodium labeling before Gore's committee in 1981. The two days of hearings that followed, while meant to focus on labeling, also served to bolster both Gore's image as a government watchdog and the idea that the science on salt was settled.[82]

Gore declared in the hearings' opening statement: "Unlike most areas concerning the relationship of diet to health, I believe that we can begin to see a consensus emerging in the research community about the reduction of salt intake for the average American."[83] To maintain this veneer of consensus, all but one of the witnesses invited to testify at the hearings were committed to the salt-hypertension hypothesis. Despite that, the testimony would reveal the flimsiness of the evidence underlying that hypothesis and just how little consensus existed among researchers.

For the appearance of balance, the committee invited a sole skeptic of the salt-hypertension hypothesis, John D. Laragh, director of the Cardiovascular Center at New York Presbyterian Hospital. After hearing testimony from the author of a low-salt cookbook, Robert Levy of NHLBI, and from Ray W. Gifford of the American Medical Association, Laragh made his best attempt to redeem salt's reputation. "There is an old saying originating in the Bible that salt is the essence of life," he testified, explaining that adequate and even surplus amounts of sodium are beneficial for survival. The idea that salt reduction would improve public welfare "arose largely from the premature and perhaps mistaken view that this would have an important bearing on high blood pressure problems," Laragh told the committee. Furthermore, he noted that because research had never considered the benefits or harms salt restriction might have in healthy individuals, it should not be recommended to the general public.[84]

Laragh also raised concerns that research, which he conceded showed a possible link between sodium intake and blood pressure, had not answered the more important question about whether lower-salt diets actually improved health outcomes. He pointed to Dahl's five-country study, which seemed to show that populations with greater salt in their diets had higher blood pressure, but these high-salt populations also tended to have better overall health. The northern Japanese, he noted "have many less heart attacks and are apparently experiencing a longer and happier life than [Americans] despite their higher sodium chloride intake." On the other hand, the tribal groups referenced throughout the committee hearings with low salt and low blood pressures had significantly reduced lifespans, which Laragh supposed was, at least in part, *because* of their low-salt diets.[85]

Displeased with Laragh's testimony, Gore noted that "one of the hazards of seeking out a balanced panel . . . is that one might inadvertently give the impression that one out of seven does view the evidence in the way Dr. Laragh views it," and accused the researcher of being in the minority within the medical community.[86] Laragh refuted that accusation and also warned that many recommendations on health for which there appeared to be consensus had later been proven erroneous. "I suggest you all look back at the fairly recent national recommendations or those of the American Heart Association which have been made, most of which have been abandoned. Thus, cholesterol or fat in the diet . . . anticoagulation for coronary disease. I can't think of any of these national advices that have survived. I wish this one better luck."[87]

The committee served not only as a platform from which to promote the idea of salt as a dangerous additive, but also as a vehicle for promoting government funding of the issue.[88] When asked if NHLBI had received adequate support to answer many of the questions posed by Laragh, Dr. Levy diplomatically noted that "the funds that we will have will allow us to acquire more knowledge. Obviously, we could learn more rapidly with more effort." Michael Jacobson, executive director of the Center for Science in the Public Interest, testified that not only should government enforce mandatory sodium labeling and restrict the level of salt in commercial foods, it should also increase its funding for public education campaigns.[89]

Gore told the committee in his closing statements: "It seems curious to me that an administration willing to spend essentially unlimited amounts of money to save a downed aviator would be willing to essentially abandon some 35 to 45 million Americans who are in a sea of salt and processed foods that pervades their diet." He further noted that "there is now a consensus in the scientific community about the relationship between sodium [and blood pressure?]." While the Gore committee could not institute binding regulations, it did provide a path forward for those both in and outside of government who wanted government to take action on salt. Its five-point plan essentially granted government approval for the subsequent anti-salt crusade on which the FDA would soon embark.[90]

Anti-Salt Propaganda

After hearing testimony, Gore's committee released a five-point plan of action on salt. This included:

- working with industry to achieve voluntary sodium reduction in foods;
- rules requiring sodium content declarations;
- possible legislative options for sodium labeling;
- public education to raise awareness of the effects of sodium on health; and
- programs to monitor the nation's sodium consumption.

The plan was sold to Secretary of Health and Human Services (HHS) Richard Schweiker as a cost-effective way to prevent disease while avoiding burdensome mandatory regulations. It was also viewed as a means of answering CSPI's petition on salt. But it would become an excuse for the newly minted FDA commissioner to undertake what can only be described as a campaign of propaganda against salt.

Arthur Hull Hayes began his tenure as FDA commissioner on the first day of the Gore committee hearings in 1981. A pharmacologist and the former director of the hypertension clinic at the Hershey Medical Center, Hayes was already a proponent of the salt-hypertension hypothesis before taking office, and reducing American's sodium intake became a centerpiece of his tenure.

In the year following the Gore committee hearings, Hayes and Schweiker held a series of meetings with industry to discuss voluntary salt reduction in food. While Schweiker told attendees he was committed to voluntary action, Hayes warned that these discussions were not open-ended and reserved the right to impose mandatory requirements if industry failed to act to his satisfaction. Those threats were backed up when Gore, Rep. Henry Waxman (D-CA), and Rep. Neal Edward Smith (D-IA) introduced a bill that would make sodium labeling mandatory.[91]

Under Hayes's direction, the FDA undertook a publicity campaign that stoked public fears over salt. He gave speeches, sent articles to newspapers, and even produced television and radio commercials advertising the importance of salt restriction. Hayes even went so far as to send letters to every physician in the United States to "raise awareness" of the role of sodium in hypertension. His behavior led at least one senior HHS official to accuse Hayes of "overstepping the bounds of scientific as well as political propriety."[92] Health organizations, including the American Medical Association, also expressed concern about the FDA's myopic new focus on salt, describing Hayes's efforts as "reminiscent of the immoderate . . . scare tactics" of the McGovern report.[93]

Still, in 1982 the American Heart Association (AHA) followed Hayes's lead, publishing a statement on diet and heart disease that included a recommendation to reduce sodium intake. Notably, the AHA's guidelines did not provide an explicit limit and conceded that "the question of whether an effort should be made to reduce the intake of sodium for the general public is an unsettled matter." Sodium restriction *might* benefit susceptible individuals, they wrote, but "obviously, the relatively high consumption of salt by the U.S. public does not cause hypertension in the majority of people."[94]

Hoping to forestall mandatory action and quell rising public fears about dietary salt stoked by Hayes's campaign, the industry in 1983 began promoting an array of no- and low-salt products. Companies marketed these new products as healthier and reminded the public that "the Surgeon General says Americans eat too much salt."[95]

But like the baby-food marketing war years earlier, the food producers' efforts did little to ease consumer fear and instead served only to fuel the growing aversion to salt as an unhealthful additive. Yet, even as the salt-hypertension hypothesis was being cemented in laypersons' minds, the debate within the scientific community about its validity intensified and began spilling out into popular media.

The Politics of Public Science

Laragh may have been the most outspoken of the critics of population-wide salt reduction, but he was far from alone. The FDA propaganda campaign and advocacy efforts by groups such as CSPI sparked both greater skepticism and interest in the salt-hypertension hypothesis. However, those advocating for universal salt reduction had learned over the previous decade that they need not prove their case in the court of science. Rather, they only needed to convince key government officials and the public. Thus, many in the research community focused their attention on maintaining the veneer of consensus as evidence mounted that cast doubt on the wisdom of salt restriction.

In a 1980 edition of *The Lancet*, John Swales published his review of the evidence on salt, concluding that the "radical" advice that everyone limit intake was based on "tenuous" data from trials of patients with hypertension. He vehemently rejected claims that even if salt restriction didn't benefit the bulk of the population it couldn't hurt, noting that "if a drug was being advocated for the treatment of hypertension its toxic effects would have to be studied in much more detail than that which has been afforded . . . dietary salt restriction."[96]

Three years later, Laragh produced his own summary of the evidence surrounding the salt "myth" in the *Annals of Internal Medicine*. Like Swales, Laragh voiced his concern about the assumption that population-wide salt reduction would have no adverse effects. "This type of speculative reasoning, however appealing, should not be the basis for a major change in public health policy. . . . Public health policy should be dictated by facts, and not by hopes or opinions," he wrote, and detailed the many documented harms of low-salt diets.[97]

In the following year, Jehoiada J. Brown and his colleagues at the internationally renowned Medical Research Council (MRC) Blood Pressure Unit in Glasgow published a review of the evidence. The authors concluded that the evidence was "insufficient for a recommendation one way or the other" when it came to salt restriction.[98] The response from the anti-salt camp, as Brown later put it, was not "properly dispassionate."[99] It could, more appropriately, be described as Simpson had observed: those committed to the salt-hypertension hypothesis defended it "like a religion" from any apostate.[100]

H. E. de Wardener, a pioneering kidney specialist at Charing Cross Hospital and prominent anti-salt advocate, wrote that the MRC's assertion that the evidence was not sufficient to promote salt restriction was not only unjustified, but also irresponsible. "Such accusations might more properly be levelled at those who [question salt reduction]" he wrote, calling it "perverse and mischievous to suggest that a modest reduction in sodium intake might occasionally be harmful." De Wardener's colleague, Graham MacGregor, refuted Brown's assessment, claiming that "there is overwhelming evidence that restriction of sodium . . . does cause substantial falls in blood pressure."[101]

Of particular concern to the anti-salt camp was that these transgressive views had captured the attention of popular media, potentially hampering efforts by those like MacGregor and de Wardener to convince the public of the need to reduce salt. "Would [they] prefer that views which are controversial should be purveyed uncritically to the public as proven facts?" Brown asked of his critics.[102] As the following years of debate would show, the answer to that question appeared to be a resounding yes.

Anti-salt advocate de Wardener called it "mischievous" of Brown and his colleagues to raise concerns about the possible risks salt restriction might pose and accused skeptics of failing to consider the totality of evidence for salt restriction's benefits, saying they themselves were guilty of systematically ignoring the mounting evidence of salt restriction's harms.[103] Between 1972 and 1980, numerous studies indicated that even moderate salt restriction had the ability to *increase* blood pressure in more than one-third of individuals, cause drops in extracellular fluid (signaling dehydration), and trigger the release of vessel-constricting hormones that, independent of their effect on blood pressure, are linked with an increased risk for stroke and death.[104]

While Brown and many others implored those in the anti-salt camp to test "the benefits and disadvantages of [salt] reduction . . . impartially," too many careers and egos had been invested in the sodium restriction policy by the time researchers began conducting the large-scale controlled trials that might shed light on these questions.[105] Government organizations from the FDA to NHLBI and the Office of the Surgeon General warned the public to reduce sodium intake, while health and research bodies like the American Heart Association and the National Academy of Sciences lent those recommendations authority. Politicians like Al Gore and advocacy groups like CSPI had also made their names fighting against the "deadly white powder."[106] As evidence began hinting that the skeptics' concerns might have merit, it was too late for some to turn back on previous pronouncements. Unable to prove their hypothesis with conclusive evidence, the anti-salt camp turned instead to a war of rhetoric.

License to Shill and the "Devil Shift"

The "devil shift" in political theory describes the process of perceiving and portraying one's adversaries in intractable debates as operating from evil motives. At first, one can excuse an opponent's failure to agree as stemming from a misunderstanding of the facts, but prolonged resistance—especially when the facts are viewed as salient and obvious—causes growing suspicion about adversaries' motives.[107]

By the 1980s, the debate had attracted powerful participants, including commercial enterprises, government agencies, and health experts. Because salt-restriction opponents were viewed as a powerful force, those in favor of salt reduction had to convey their position with certitude and tactfully avoid conversations about the limitations of the evidence. Furthermore, since they maintained that the evidence was incontrovertibly in support of salt reduction, they began to characterize the persistence of the naysayers as stemming from improper motivations. The most common explanation offered was the perverse influence of industry.

In their book *Salt, Diet and Health*, MacGregor and de Wardener dedicated an entire chapter to the "industrial conspiracy," which they viewed as manufacturing the current debate about salt. Those researchers, like Brown, who remained skeptical of salt reduction were considered part of the "the pro-salt lobby," products of a "surreptitious and covert public relations campaign" funded and fomented by commercial interests.[108]

MacGregor and de Wardener targeted John Swales, the founder of the journal *Hypertension*. Though not funded by industry, Swales was criticized for participating in industry-sponsored events and for presenting his skeptical views on salt reduction at a press conference organized by the Salt Manufacturers Association, a lobbying organization for the industry. In his defense, Swales pointed out that many researchers, including MacGregor and de Wardener, participated in industry events and accepted paid consulting positions to advise these commercial interests. Such relationships, he wrote, are valid criticisms only "if there is a claim that a scientific view is being influenced by payment."[109] That was MacGregor and de Wardener's implication, though they avoided explicit accusations that would require them to provide evidence they did not have.

Anti-salt advocates made similar implications of impropriety against David A. McCarron, head of the Hypertension Program at the Oregon Health Sciences University, in 1984 after the publication of a study he conducted that showed diets deficient in certain nutrients were linked to hypertension. McCarron and his colleagues analyzed the diets and health of more than 10,000 Americans from data collected by the National Center for Health Statistics. Analysis of the data found that higher blood pressure was associated with *lower* levels of calcium, potassium, and certain vitamins. Furthermore, they found that individuals with higher intakes of calcium, potassium, and salt tended to have *lower* blood pressure.[110] The results incensed anti-salt advocates like CSPI, which publicly alleged McCarron's work could not be trusted because he had previous funding from the dairy industry—although the study in question received no industry funding.[111]

Government experts, both at the National Center for Health Statistics and the FDA, also defended the salt hypothesis with the "totality of evidence" argument. Even if industry funding had not influenced studies, they asserted that the study was misleading because it failed to put the results in the context of the "abundance of . . . data suggesting that dietary sodium indeed plays an important role in hypertension."[112] NHLBI also responded to the study, arguing that the notion that high blood pressure might be caused by dietary deficiencies rather than excesses "flies in the face of tons of information that says the reverse."[113]

It is likely that NHLBI felt compelled to defend the salt-hypertension hypothesis, as the FDA was on the verge of funding one of the largest observational studies conducted at the time, hoping to show a conclusive link between salt intake and blood pressure.[114]

The strategy of *implied* corruption would become increasingly popular among the anti-salt camp. For example, in his book *The Politics of Food*, self-described "food campaigner" Geoffrey Cannon admitted that it would be wrong to malign a researcher's objectivity simply because he or she worked for the industry. Yet that seemed to be the purpose of Cannon's detailing the case of Dr. Ian Robertson, one of Brown's MRC colleagues and coauthor of the skeptical paper. Robertson had been forced to resign his post at MRC for failing to disclose paid consulting work he had done on behalf of the pharmaceutical industry. Cannon did not provide evidence that Robertson's connection to the pharmaceutical industry influenced his work, but he pointed to the skeptical salt paper as the reason the Royal College of Physicians shelved plans to recommend salt reduction to the UK public. Though there was no evidence that Robertson's views were influenced by his connection to the industry (as opposed to the reverse), Cannon wrote that this would have "delighted" the pharmaceutical industry, which had a "vested interest in high blood pressure; the more people with high blood pressure, the more customers to carry on with [antihypertensive drugs.]"[115]

It was against this devil shift that journalist Gary Taubes published his provocative and award-winning article on the debate for the journal *Science*. After interviewing some 80 researchers, clinicians, and administrators involved in salt policy, Taubes concluded that its fever emanated from the fact that the evidence for salt reduction was so tenuous; it generated the perceived need for consensus among those researchers who advocated it. Furthermore, protecting the appearance of consensus demanded that controversy be either dismissed or explained away as the product of a profit-motivated lobby.

Taubes was warned by many interviewees that even publishing his article would play into the salt lobby's hands. Edward Roccella, director of the National Heart, Lung, and Blood Institute's High Blood Pressure Education Program (NHBPE), warned Taubes that writing about the scientific controversy served "only to undermine the public health of the nation" while Jeff Cutler, director of the Division of Clinical Applications and Interventions at NIH, commented that "as long as there are things in the media that say the salt controversy continues . . . [the salt lobby] wins."[116]

Other prominent researchers shared the opinion that the salt controversy had been manufactured by the industry through an "expensive and largely successful public relations campaign," which had convinced governments not to take action on salt.[117]

To counter this industry influence, they formed their own public relations and lobbying group. Led by MacGregor and de Wardener, Consensus Action on Salt and Hypertension (CASH) was established in the UK in 1996. In 2005, Roccella, Cutler, Stamler, and others founded a sister organization, World Action on Salt and Health (WASH). Both CASH and WASH seek to develop "consensus" and convince the industry and governments of the need to take action on salt. Their funding came primarily from the food industry—even as its members attacked their adversaries for industry ties—and, at least for CASH, from the British government. The Food Standards Agency, which CASH regularly lobbies for stricter salt regulations, reportedly provided about 12 percent of the organization's funding. In other words, the UK government is spending taxpayer money to lobby itself.[118]

None of the researchers at these organizations receive personal income from the government. However, many of those involved hold employment positions within the organizations that not only decide on national and international health policy, but also fund all of the relevant research. For example, a founding member of CASH, Francesco Cappuccio, served as director of the Collaborating Centre for Nutrition at the World Health Organization (WHO). Another WASH affiliate, Caitlin Boon, is a food and nutrition board program officer at the

Institute of Medicine (IOM). Numerous affiliates of NHLBI have participated in CASH or WASH, including NHLBI director Claude Lenfant, program director Winnie Barouch, and outspoken salt reduction advocate Paul K. Whelton.

Whelton, who conducted landmark hypertension trials, served as co-chair of NHLBI's advisory panel on hypertension prevention, part of NHLBI's National High Blood Pressure Education Program. Half the members of this panel, along with Whelton, were WASH affiliates, including one of its most influential members, Lawrence J. Appel, who chaired the IOM's Dietary Reference Intakes for electrolytes and water when the institute first made its recommendation that adults limit sodium intake to 2,300 mg, and also served on two Dietary Guidelines Advisory Committees.[119]

With the considerable influence these consensus builders wielded, it should not be surprising that even as the evidence began indicating that salt did not have a primary role in the development of hypertension, and that salt reduction could cause severe harm to public health, government agencies have only increased their efforts to spread the salt dogma.

Modern Research and the Undoing of the Salt Myth

While the consensus builders insisted that the science on salt was settled, both supporters and critics of this supposed consensus recognized that the existing evidence was rather weak. Since the basis for the salt-hypertension hypothesis arose mainly from poorly controlled observational studies and a few clinical trials of the ill, by the 1990s researchers endeavored to conduct the large and better-constructed studies they hoped would finally prove that dietary salt affected the blood pressure of the general population and which would thus quiet the naysayers.

A Scottish heart health study published in 1988 was, at the time, the largest population study investigating salt in the diet. After collecting data from 7,300 Scottish adults on 27 different lifestyle factors, Hugh Tunstall-Pedoe and his colleagues at the Ninewells Hospital concluded that while higher potassium intake seemed related to lower pressure, the association with salt was "weak and does not have any real independent role in explaining blood pressure."[120]

These results, in a way, affirmed what McCarron had found years earlier. Yet the Scottish Heart study received little attention. Critics dismissed it for its "methodological problems" and instead turned their focus toward a different study: the NHLBI-funded Intersalt, which was published in the same issue of the *British Medical Journal*. Led by Jeremiah Stamler, Intersalt was the largest study of its kind at the time, and while some hailed it as the nail-in-the-coffin of the salt-hypertension hypothesis skeptics, it soon proved as divisive and polarizing as those that had come before it. Both sides of the salt war interpreted the study as supporting their case, and the subsequent battle over whose interpretation was correct would serve to highlight the fundamental errors underlying the entire salt debate.

Intersalt I

The Intersalt Cooperative Research Group, a consortium of nearly 150 researchers around the world, gathered data from more than 10,000 adults, randomly selected from 52 sites. When the researchers analyzed the data, accounting for confounding factors like alcohol intake and weight, the team found a positive, but weak, connection between salt intake (estimated by 24-hour urine samples) and blood pressure that seemed to show that for every 2,300 mg reduction in daily sodium intake there was a corresponding reduction in blood pressure of about 2.2 points systolic and one point diastolic.

The results seemed to confirm a link between salt intake and blood pressure, but the differences were "so low even the smallest harmful effect could negate or reverse the advantage of such a reduction in blood pressure," Swales wrote in his accompanying editorial.[121]

Furthermore, of the 52 data collection sites, four were set among tribal societies—the Yanomamo and Xingu tribes in Brazil, natives of Papua New Guinea, and tribal Kenyans. These isolated populations had extraordinarily little sodium in their daily diet, some less than 100 mg a day. While they did have low blood pressure, their diets and lifestyles also differed from those in industrialized populations in just about every other conceivable way. Once the data were analyzed excluding these four outlier populations, the weak relationship the authors found between salt and blood pressure disappeared.

The most impressive result obtained by Intersalt was the connection it seemed to show between salt and blood pressure increases with age. The study's authors asserted that reducing sodium intake by about 2,300 mg a day would cause blood pressure rises between age 25 and 55 to be reduced by about 9 points systolic and 4.5 points diastolic.[122]

Intersalt's lead authors promoted their findings as "abundant, rich, and precise confirmation" of the benefits of, and need for, salt restriction in the general population.[123] Some in the media carried that message, but most in the scientific community viewed Intersalt as evidence against the salt-hypertension hypothesis. As William Bennett, editor of the *Harvard Medical School Health Letter*, lamented, the study had provided the best data researchers could hope for in the next century, and what it showed was that "for all practical purposes, salt is a minor factor in the development of hypertension, once consumption exceeds a very low level."[124]

Intersalt's findings were so lackluster that even some supporters of the salt-hypertension hypothesis began backing away. A November 1990 Associated Press story headlined "For 90 Percent of Americans, Salt Doesn't Matter Much" quoted various health experts disappointed by Intersalt. John Vanderveen, director of nutrition at the FDA, admitted that there was "no conclusive evidence that salt consumption causes hypertension. It's only a hypothesis." John C. LaRosa, chairman of the AHA's Nutrition Committee, commented, "You make these recommendations and the science changes and you have to be able to back away from them. . . . You've got to do that in such a way that you don't destroy your credibility."[125]

Even a few of Intersalt's researchers, such as Lennart Hansson of the University of Uppsala, viewed their study as failing to confirm salt's link to blood pressure. "It did not show blood pressure increases if you eat a lot of salt," Hansson—a pioneer in research on anti-hypertensive therapies—later told reporters. As Swales put it, Intersalt's findings "hardly seem likely to take nutritionists to the barricades except perhaps the ones already there." Yet many of those already on the barricades held influential positions in the government agencies that controlled not only the narrative on salt, but also the purse strings for scientific research.[126]

Government Support for Government Advice

The National Heart, Lung, and Blood Institute, which funded Intersalt (and funds most hypertension research in the United States), echoed the interpretation of lead author Jeremiah Stamler, calling the study, in a 1993 report, confirmation of a "strong positive relationship" between dietary salt and blood pressure. Unsurprisingly, the chairman and half the members of the working group that developed this report were researchers affiliated with WASH.[127]

NHLBI, which had spent more than $1 million funding Intersalt, continued to support Stamler's work in the following years, allowing the authors to go back to their data and reanalyze their original findings using different methodologies. "Intersalt Revisited," published in 1996, no longer found a "weak" connection between salt and blood pressure, but a strong and clear link—a threefold increase in the blood pressure benefits of reducing sodium intake.[128]

Intersalt Reengineered

Intersalt Revisited eliminated correction for body mass and added a controversial correction for something known as regression dilution bias. In short, regression dilution bias refers to the tendency for small errors in measurement (e.g., estimates of sodium intake) to weaken the strength of a relationship between variables (e.g., between sodium and blood pressure). The authors assumed that by basing sodium intake on 24-hour urine samples, which do not necessarily reflect an individual's overall intake because intake varies from day to day, their estimates would be slightly inaccurate. In guessing by how much their estimates would be off and accounting for that, the authors hoped to eliminate the presumed dilution bias and strengthen the assumed association between salt and blood pressure. Of course, adjusting data based on assumptions in this way can have the effect of magnifying a relationship that isn't there to begin with.[129] Still, once its authors adjusted their data in this manner, Intersalt Revisited was able to report a strong and positive connection between sodium and blood pressure.[130]

The anti-salt camp ran with the new results. Researchers like Malcolm Law of the Medical College of St. Bartholomew's Hospital in

London and a founding member of CASH, heralded Intersalt Revisited as incontrovertible confirmation of "salt as an important determinant of blood pressure."[131] Others, however, viewed the reworked study with skepticism. George Davey Smith, a clinical epidemiologist at the University of Bristol, called the correction for regression dilution bias "misleading," and Nicholas E. Day, director of the MRC Biostatistics Unit at the Institute of Public Health in Cambridge, declared that "statistical complexity should not be used to conceal inadequacies of the data." James Le Fanu of the Mawbey Brough Health Centre in London simply questioned "why the combined intellects of so many distinguished epidemiologists should maintain that the evidence incriminating salt in hypertension is so convincing when clearly it adds up to very little."[132]

But the most damning criticism of Intersalt, and of the entire scientific discussion around salt, was the fact that none of these studies—even if they showed a small connection between blood pressure and sodium—answered the more important question: Does it matter for health?

Salt, Blood Pressure, and Mortality

Hypertension is not, in and of itself, a negative health outcome. That is, nobody dies from high blood pressure alone, but rather from the diseases and conditions closely linked to high blood pressure, like cardiovascular disease and strokes. Furthermore, merely reducing blood pressure, as studies of hypertensive drugs have found, is not necessarily enough to reduce the risk of death.[133] But all of the evidence on which the sodium reduction hypothesis was based investigated only sodium's effect on blood pressure.

Although proving the existence of a link between salt intake and a health marker like blood pressure is important for future research, it does not provide insight into how alterations of behaviors and markers affect health *outcomes*. Populations with high-salt diets, even when they have higher average blood pressure, do not necessarily have worse health. Take, for example, the Japanese, which as Dahl found in his population studies had some of the highest salt levels of any group recorded. The Japanese, however, have fewer heart attacks and live longer than people in cultures with lower-salt diets and lower blood pressure.

This lack of *outcome* evidence had long disturbed many researchers engaged in the salt debate. One of them was Bjorn Folkow of Gothenburg University in Sweden, a universally respected leader in hypertension research. Until Intersalt, Folkow had remained tactfully neutral, noting that both sides had merit, though both seemed to make their points by "overemphasizing their favourite angle of an indeed complex problem."[134] After Intersalt, however, Folkow began raising the alarm that the research community was ignoring fundamentally important questions.

In a 1998 paper, Folkow and a colleague, Daniel Ely of the Department of Experimental Biology at the University of Akron, argued that blood pressure, on which much of the debate had focused, was not actually the best measurement of heart health.[135] Research into heart health had, they asserted, repeatedly confirmed that *heart rate*, preferably heart rate combined with systolic or mean blood pressure, was a far more accurate indicator of cardiovascular stress and thus a better predictor of negative cardiovascular outcomes. As Folkow and Ely note, clinical research in both animals and humans had shown that, as salt is increased in the diet (within the normal range), heart rate actually decreases. Furthermore, when salt is restricted, heart rate goes up and rises much more than systolic blood pressure declines. That is, rather than benefiting the general public, salt restriction actually seemed to increase cardiovascular stress.

Despite knowledge of this information, research continued to focus only on blood pressure. Folkow hinted that the reason was that those conducting the research knew that publishing data on the effects of salt restriction on heart rate would harm their efforts to change public policy. He pointed specifically to the Intersalt study, which, according to one of its authors, Paul Elliot (a founding member of CASH), *had* collected data on heart rate, but had never included it in any of its analyses or made it public. The authors claim the reason they declined to release their underlying data was to maintain the investigation's "independence."[136] That might be the case, but as Folkow and Ely pointed out, it's also likely they found that lower salt diets are associated with higher heart rates—a result that would have led to questions about the overall benefits of sodium reduction.[137]

The connection between salt reduction and elevated heart rate, as well as between heart rate and mortality, has been repeatedly confirmed by clinical data.[138] These warning signs did little to slow the anti-salt camp's efforts, but they did pique the research community's interest. By the 2000s, the question of how salt reduction affects health became a focus of scientific investigation. In addition to escalating the skeptics' concerns, this modern era of research increasingly indicated that other dietary and lifestyle factors might be more important for blood pressure and health than salt.

The Era of Meta-Analysis

There are many theories as to why 95 percent of the world's population consumes more than the recommended level of sodium; researchers simply don't have an answer. Without understanding what drives individuals to consume a particular level of sodium and what benefits that level might confer, it is impossible to know what harm might be done by forcing them to lower their intake below that level.[139] Some, like Pavel Goldstein and Micah Leshem of the University of Haifa, have theorized that higher salt levels may be important for periods of growth (e.g., adolescence and pregnancy) and that increasing salt may somehow counteract the effects of stress. For example, in a study they found that dietary sodium was inversely associated with depression in women.[140]

While world governments forged ahead with programs to reduce salt intake, spurred on by anti-salt activists telling them it would save lives, it wasn't until the 2000s that researchers really began investigating the effect salt restriction might have on health and risk of death. What many found was that, far from seemingly reducing risks, low-sodium diets actually appeared to *increase* risk of death in the general public. But as is typical for the salt debate, others came to opposite conclusions—some even when looking at the *same data.*

For example, in 2008 a team let by Hillel Cohen of the Albert Einstein College of Medicine examined data from the third National Health and Nutrition Examination Survey, which, based on a single 24-hour dietary recall (in which participants recall all they ate in the previous day) contained estimated sodium, potassium, and calorie intake for nearly 9,000 people recruited between 1988 and 1994. After adjusting for known cardiovascular disease risks (e.g., weight, smoking, serum cholesterol) and confounding dietary factors like potassium intake, the researchers found that people consuming the least sodium (under 2,060 mg a day) were more likely to die from cardiovascular disease than those with a daily sodium intake between 4,000 and 10,000 mg.[141] In response, researchers at the U.S. Centers for Disease Control and Prevention (CDC) quickly reanalyzed the data and concluded just the opposite, finding that increased sodium intake was linearly associated with all-cause mortality.[142]

As Cohen and his coauthors later charged, the reason CDC was able to find a result contrary to their own was because the CDC researchers—unlike Cohen's team—did not adjust their data for potassium intake. Instead, the CDC researchers adjusted the data for the sodium-to-potassium ratio (the difference between intake of the two nutrients). Unlike for salt, the data on potassium are unequivocal: it has long been known that higher intakes of potassium are strongly associated with both lower blood pressure and reduced mortality. Thus, by using a ratio of sodium to potassium, rather than potassium alone, CDC allowed lower potassium intakes to potentially drive the relationship they found between higher sodium and mortality.[143]

In 2011, Katarzyna Stolarz-Skrzypek and her colleagues at the University of Leuven in Belgium conducted one of the first prospective observational studies to look at health outcomes and salt intake. After estimating sodium intake for 3,600 individuals, via single 24-hour urine samples, the team tracked the group's health for an average of eight years. They found that, although greater sodium intake was associated with higher blood pressure, "this association did not translate into a higher risk of hypertension or [cardiovascular disease complications]." In fact, they found that those with lower sodium intake had a *higher* risk of cardiovascular death.[144]

It was perhaps a shock for the anti-salt camp that Stolarz-Skrzypek, a member of WASH, had published a study so damaging to their campaign, and it provoked an onslaught of criticism. Predictably, the always-vocal CASH members, including MacGregor and de Wardener, spoke out against the validity of Stolarz-Skrzypek's study's methodology. For example, they argued that the method of urine collection had biased the results—even though the methodology was virtually identical to Intersalt's—and thus her study "cannot be used to refute the strong evidence that a modest reduction in population salt intake is extremely beneficial to health and could prevent millions of deaths."[145] Remarkably, *The Lancet* published an unsigned editorial about the *Journal of the American Medical Association* study, calling it "disappointingly weak . . . [and] likely to confuse public perceptions of the importance of salt as a risk factor for high blood pressure, heart disease, and stroke."[146] However, subsequent research investigating the relationship between dietary salt and death confirmed Stolarz-Skrzypek and her colleagues' findings.

In the same year, Martin O'Donnell of the Population Health Research Institute at McMaster University and his colleagues analyzed data for nearly 29,000 people participating in heart disease drug trials. They found that, within this high-risk population, cardiovascular death was highest among those with the highest and lowest salt intake. This U-shaped curve showed that subjects consuming an average sodium intake of between 3,000 mg and 7,000 mg appeared to be at the lowest risk for death.[147] They found similar results in the general population three years later, when, as part of the Prospective Urban Rural Epidemiology (PURE) study, O'Donnell and colleagues analyzed data for more than 100,000 healthy adults in 17 different countries. They found that those consuming a diet of more than 7,000 mg of sodium a day (estimated via single 24-hour urine samples) appeared to have a 15 percent higher risk of death. But they also found that those with diets containing less than 3,000 mg of sodium a day had a 27 percent higher risk of death.[148]

About a month later, Graudal and his colleagues confirmed this higher risk of death from low salt in their 2014 meta-analysis, which pooled data from 27 different studies. The data, pulled from cohort and randomized controlled trials, indicated that risk of death was lowest among those consuming between 2,645 mg and 4,945 mg of sodium a day.[149] CASH members MacGregor and Feng J. He, a senior research fellow at Queen Mary University of London's Wolfson Institute of Preventive Medicine, argued that the study had methodological flaws, pointed to the "totality" of evidence against salt, and again raised the specter of industry influence by accusing Graudal's coauthor, Michael Alderman, of having a conflict of interest because he had previously worked as a paid industry consultant.[150]

Two years later, in the 2016 follow-up to the PURE study, O'Donnell and his team again found that among the 130,000 individuals studied—half with hypertension, half without—sodium intake below 3,000 mg a day was associated with increased mortality risk in normotensive people.[151]

Before his death, Robert P. Heaney, an endocrinologist at Creighton University, noted that these studies confirmed what researchers should have been able to guess based on their understanding of all other essential dietary elements. It is "reassuring, to note that sodium is thus like most other nutrients in that there is potential harm at both extremes of intake," he wrote, commenting also that attempts to lower population salt intake from the average "cannot be defended by available evidence."[152]

The anti-salt camp responded with increasing hostility to each new study showing potential harm from universal salt restriction, at times even attacking publications for accepting such studies. *The Lancet*, for example, was lambasted for publishing O'Donnell's 2016 PURE results. Francesco Cappuccio, a CASH member and head of the World Health Organization's Collaborating Centre for Nutrition, wrote of his "disbelief" that the journal would give a platform to such "bad science."[153]

As Graudal later pointed out, there had not been a single randomized controlled study to examine how reducing sodium to the government's recommended limit of 2,300 mg per day might affect the health of the general population.[154] Despite this, some health agencies, including the WHO, which shared members with CASH and WASH, decided to continue to make salt reduction a priority. As MacGregor put it in 2014, they did so because salt reduction is "cost-effective to implement and so easy to do."[155] That assertion is highly debatable, given the experience most nations have had with salt-reduction attempts.

Salt Reduction "Successes"

The United Kingdom is often cited for its success with salt. Beginning around 2003, the Food Standards Agency, along with CASH and members of the food industry, began a comprehensive program to lower salt intake. This included both educating the public about the harms of too much salt and convincing food manufacturers to lower the overall sodium content of commercial foods. By 2009, MacGregor and He reported success from the program, noting a 20–30 percent reduction in salt in foods and a corresponding drop in salt intake of about 15 percent. According to government data (based on 24-hour urine sampling), by 2008 the average salt intake of UK residents had dropped from 9.5 grams per day to about 8.6 grams, blood pressure fell by an average of three points systolic and 1.5 points diastolic, and death from stroke and ischemic heart disease had declined by 42 and 40 percent, respectively. While MacGregor and He admitted that many other factors likely contributed to the lower death rate (e.g., less smoking), they concluded that the observed reduction in salt intake "would have contributed substantially to the decrease in stroke and IHD mortality."[156]

The actual amount of salt reduction has been disputed, but it is certainly possible that salt consumption in the United Kingdom declined by the modest 10 percent (about 400 mg sodium a day) reported by MacGregor and He.[157] At the start of the program, sodium intake was an average of 3,700 mg per day and declined to 3,300 mg per day by the end. The "low" level achieved by the UK puts it not only on par with average U.S. intake, but also well within the normal range observed around the world. Setting aside the question of whether governments *should* try to lower UK salt intake, it is questionable—based on the experience of other nations—whether they *could* lower intake.

Finland has arguably had the greatest success of any nation with salt reduction. Beginning in the 1970s, the Finnish government initiated an ambitious program aimed at reducing heart-related diseases by lowering intake of salt and alcohol, reducing smoking and weight, and increasing exercise. Over the course of 40 years, the nation managed to bring its average salt intake down from a high of 14 grams

(5,400 mg sodium) to around 8 grams (3,100 mg sodium) a day in 2012. Although this decline is impressive, it too is within the normal human range. Furthermore, the drops, which were rapid at first, slowed as intake approached the lower end of this normal range, and in some areas of the nation salt intake has slightly ticked upward.[158] It is also worth noting that while anti-salt activists point to Finland's sodium reduction as the cause of the nation's increased life expectancy and improved health, the United States has had even more significant reductions in cardiovascular death during this time period despite no change in salt intake.[159]

Finland's experience indicates that salt reduction programs might be able to modify average intake, but only within the relatively narrow range identified by McCarron, Graudal, and others (between 3,000 and 6,000 mg of sodium per day).[160] Getting people to go above or below this range—toward the target of six grams of salt (2,300 mg sodium)—might be far more difficult. This is something researchers have frequently reported in sodium restriction trials, noting the difficulty of getting participants' intake lower than the average. For example, the second phase of the Trials of Hypertension Prevention, known as TOHP-II, set a goal of less than 1,840 mg sodium for participants. Even with intensive counseling, the average intake managed only to reach between 2,390 mg and 2,852 mg per day. Furthermore, this average crept back up to 3,200 mg a day by the 36th month of the trial.[161] In the Trial of Nonpharmacologic Interventions in the Elderly that examined weight loss and sodium reduction in elderly hypertensives, the researchers set a target goal of less than 1,800 mg sodium a day, but mean intake fell only to an average of 2,400 mg per day. In the follow-up, patients in the sodium reduction arm of the study had far more adverse cardiovascular events than those in the weight reduction or even weight- and sodium-reduction groups (and only slightly fewer events than the control group).[162]

Recent research provides insight into why we are compelled to maintain a certain "high" level of sodium intake. As salt in the diet is decreased, it triggers powerful compensatory mechanisms, designed to reestablish homeostasis—to keep body fluids and salts stable. In particular, low sodium or low fluid volume triggers something known as the renin-angiotensin-aldosterone system (RAAS), which stimulates thirst, sodium and water retention, and vessel constriction to elevate blood pressure. For example, very low salt cultures, like the Yonomamo Indians, compensate for their lack of sodium with the extreme production of sodium-retention hormones renin and aldosterone.[163] However, elevated levels of the RAAS hormones are also associated with cardiovascular death, independent of their effect on blood pressure.[164] These compensatory mechanisms may explain, in part, why low-sodium groups like the Yonomamo have a shorter life expectancy (in their case, a mean survival age of 40 years) and why people around the world find reducing sodium intake below 3,000 mg per day so difficult.[165]

In 2011, the year before Dr. Folkow died at age 91, he warned his fellow researchers of the danger of attempting to "overrule" the body's desired level of sodium by political decree. He argued that those continuing to promote universal sodium reduction, without taking into account the evidence of potential harm he and others had noted, violated the fundamental tenets of science. As an example, Folkow pointed to a report from the Institute of Medicine as the epitome of the scientific bias he had observed throughout the salt debate.

Issued in 2011, the IOM report identified hypertension as one of its top health priorities and identified sodium reduction as a primary recommended approach. Among the references cited in support of IOM's conclusions, Folkow noted "a striking lack of mention of some . . . really important data studies or reviews, which do *not* consider [current average salt intakes] as any real health threat. . . . It is instead concluded that daily [salt] intakes in adults should be lowered to 6 g—according to some in the panel to 3–4 g. But little mention about how this could be achieved, and whether it in the end would have much effect."[166]

Folkow did not point out that the lead reviewer of this IOM report was none other than Lawrence J. Appel, a professor of medicine at the Johns Hopkins University's School of Public Health, who had long involved himself in the salt-hypertension issue, serving on numerous government advisory panels and sitting on the board of WASH. Appel chaired IOM's panel on Dietary Reference Intakes for electrolytes and water when the institute first recommended that adults limit sodium intake to 2,300 mg per day. He also served as a member of the Dietary Guidelines Advisory Committee, which in 1980 had noted that salt restriction was advisable for those with high blood pressure and that "not everyone is equally susceptible." But the 2005 guidelines—with Appel's input—contained none of these caveats. Instead, they followed IOM's lead, recommending that most Americans limit intake of sodium to no more than 2,300 mg a day and the rest (African Americans and those over 51 years old) limit intake to even less, 1,500 mg per day.[167]

The quantitative sodium limit settled on by the government, and ultimately all health agencies, was based on estimates of average daily sodium loss. The limit does not account either for people living in hotter or more humid climates or for greater losses that might occur with greater levels of physical activity. Since Folkow's death, government bodies have continued to recommend this limit, almost universally failing to warn people that evidence has shown that, under nonaverage conditions, it is too low and can cause acute harm.[168]

In addition, they continue to ignore evidence that, for the average person, this limit is not only unhelpful but also makes it impossible to get adequate amounts of other vital nutrients like potassium.[169] Instead, they continue to reject any data linking their recommended sodium limit with greater risk of death and worsening health markers such as the increases in plasma renin, aldosterone, noradrenaline, adrenaline, cholesterol and triglycerides, observed when people shift to low-salt diets.[170] They argue that there is a totality of evidence, as well as a consensus on the benefits of universal salt reduction, and that the debate is (or ought to be) over.[171]

In other words, while scientific understanding of sodium processing, blood pressure and fluid regulation, and the impacts of the central and sympathetic nervous system on these factors had advanced enormously in the preceding decades, the behavior and message of anti-salt advocates remained remarkably constant. In fact, as evidence in favor of universal sodium reduction weakened, anti-salt advocates somehow became more certain and intense in their efforts.

In 2016, Ludovic Trinquart of Columbia University and his colleagues studied the 36-year debate and found roughly equal evidence on both sides of the debate. "The divide between the uncertainty in the scientific literature about the potential benefits of salt reduction in populations and the certitude expressed by decision makers involved in developing public health policies in this arena is jarring," the authors observed.[172] While much attention is given to financial conflicts of interest, their findings indicated that the scientific debate over salt might be more biased by sheer commitment to the status quo.

The need to defend the prevailing wisdom is reinforced not only by individual psychology (e.g., confirmation bias) but also through social pressure (e.g., the efforts of CASH/WASH) and perverse incentive structures created by government involvement in science, where challenging the status quo may present risks to an individual's career. As psychologist Michael E. Oakes wrote, the manner in which the salt debate has progressed is "a testimony to the lack of objectivity of science in practice. . . . When all else fails some salt scientists call the perceived rival a shameless carpetbagger and resort to the use of complex and controversial statistical procedures . . . to conceal and confuse and interpretations of results are made cockeyed so that they are in harmony with a longstanding agenda."[173]

We may never know exactly how efforts to reduce our salt intake have harmed our health, but one opportunity cost is clear. When health authorities chose salt as the main modifiable risk for hypertension, they directed attention and resources away from other approaches. Such a decision influences not only how consumers tailor their diets, what products industry sells us, and what programs government funds, but also what questions researchers choose to investigate. In other words, it influences the progression of the science of nutrition itself. Pressure to continue the campaign to reduce sodium intake has thus overshadowed other strategies that might be safer and more effective at promoting health.[174]

Can Cooler Heads Prevail?

The history of dietary advice in America is largely a history of repeated failure. Many still remember the *Time* magazine cover from 1984 featuring two fried eggs and a strip of bacon on a plate in the shape of a frowning face. The accompanying headline, "Hold the Eggs and Butter; Cholesterol Is Proved Deadly," helped spread the dogma that diets high in fat and cholesterol almost invariably lead to heart disease.[175]

Yet, after 15 years of bland breakfasts, *Time* featured a nearly identical cover, but with a slice of melon in place of the bacon and the face now smiling. Cholesterol, the magazine reported, is no longer something to worry about.[176] The reversal on cholesterol came as a result of emerging research that the original warnings were not—and had never been—supported by science. But it would take the U.S. government an additional 16 years to update its own guidelines, finally forced to admit in 2015 that cholesterol was no longer a nutrient of concern.

Similar to cholesterol and salt, dietary fat came to its place of villainy thanks to outspoken experts willing to make logical leaps based on early studies. Like Dahl, these early studies seemed to show that populations with more fat in their diet had higher rates of heart disease. Like the recommendations to reduce salt, universal fat-reduction recommendations in the initial dietary guidelines were based on the unjustified assumption that lowering its intake would improve public health. Instead, as Americans ditched full-fat foods for low- and no-fat alternatives, we traded fats for carbohydrates and, perhaps not coincidentally, experienced a massive surge in rates of diabetes and obesity. And like the salt debate, the debate around fat was fierce from the beginning, and the case against fat became flimsier over time, as research indicated that full-fat foods, rather than causing chronic disease, might actually offer protection *against* disease like diabetes and obesity.[177] But it would take the U.S. government decades to reverse itself, only admitting in 2015 that the focus should be on "optimizing types of dietary fat and not reducing total fat."[178]

Saturated fat, once labeled a "poison" by groups like CSPI, which urged consumers to replace foods such as butter with "healthier" trans-fat-laden foods like margarine, now seemed poised for a similar reversal.[179] Government advice continues to warn that saturated fat ought to be limited, and preferably replaced with polyunsaturated fats. But in recent years, saturated fats have been found to convey different risks and benefits, depending on their type. For example, full-fat yogurt, a single serving of which might contain up to 20 percent of the daily recommended limit of saturated fat, has been shown not to be associated with higher rates of heart disease or diabetes and is actually associated with lower risk of obesity.[180] Yet, while the *2015 Dietary Guidelines* did not do away entirely with the recommendation to limit saturated fat, it did admit, "Cutting back on saturated fat will likely have no benefit." Health authorities deserve credit for recognizing the errors of their predecessors and trying to bring dietary recommendations more in accord with the actual science. But they appear not to have learned from these previous mistakes. One-size-fits-all guidelines for the American public ought to be made with extreme caution and ought to be based on exceedingly strong scientific evidence, rather than on the exaggerated pronouncements of strong-willed "experts." Furthermore, such recommendations should admit to the limitations of the underlying science and be accompanied by appropriate caveats about potential risks. The continuing failure to adhere to these standards threatens to erode public trust in government dietary guidelines and health advice entirely. As Laragh and many others warned in the 1970s, this erosion of trust not only harms efforts to educate the public about nutrition, but may also influence other public health initiatives, such as the management of contagious diseases through vaccination.

Rather than sticking stubbornly to the increasingly uncertain dogma of salt reduction, health authorities should pivot toward hypertension prevention strategies that are supported by better evidence, while some measure of public trust in health advice remains intact. Unlike the research on salt, the evidence supporting weight loss, increased potassium intake, and increased consumption of fruits and vegetables truly is unequivocal. But the fixation on salt reduction has left too many people unaware that these other strategies, in addition to improving other aspects of health, may be more effective for lowering blood pressure, preventing hypertension, and improving heart health.

Potassium

As the wisdom of universal salt reduction has grown increasingly questionable, evidence for the ability of increased potassium to lower blood pressure has only grown stronger. Though anti-salt activists railed against McCarron for his 1984 study suggesting that deficiencies in certain nutrients like potassium might be more to blame for high blood pressure than sodium excesses, there is now practically universal accord among researchers, even those who disagree on salt, that higher potassium consumption is associated with lower blood pressure.[181]

Unlike Dahl's attempt to find a difference in sodium intake between normotensive and hypertensive populations, studies have consistently shown that hypertensive individuals have lower potassium intake than their normotensive counterparts.[182] Although the Scottish Heart Health Study and Intersalt came to opposite conclusions on salt, both reported a significant inverse association between potassium and blood pressure.

The association between potassium and health outcomes is similarly consistent. A 2009 meta-analysis of 13 studies by Pasquale Strazzullo and colleagues at the University of Naples Medical School reported "unequivocally that higher salt intake is associated with a greater incidence of strokes and total cardiovascular events."[183] The study has been criticized for excluding data from the analysis that was unsupportive of the authors' preferred conclusion, but it remains a regularly cited piece of evidence in support of sodium reduction.[184] However, Strazzullo and others continue to ignore that of the five studies included that looked at potassium intake, all found that higher potassium intake was strongly and inversely associated with negative health outcomes, in some cases more strongly than sodium intake.[185]

Similarly, while O'Donnell's PURE study from 2014 found a nonlinear relationship between sodium and health (a U-shaped curve with greatest risk at the highest and lowest ends of intake), it also found an inverse association between higher potassium and mortality.[186] Nancy Cook and her colleagues at the Harvard Medical School contested O'Donnell's findings and conducted their own study, which found a "non-significant" linear association with higher sodium (not the U-shaped curve O'Donnell reported). Yet Cook and colleagues confirmed O'Donnell's potassium finding, noting that the sodium-to-potassium ratio was an even stronger predictor of health outcomes than either sodium or potassium alone. Cook's 2016 follow-up found no significant difference in mortality rates between groups on various levels of sodium intake, but it did find a significant relationship between mortality and the sodium-to-potassium ratio.[187]

The connection between potassium and blood pressure has been observed in every subgroup of the population, including the elderly, African Americans, and even normotensive individuals.[188] A reasonable criticism of this theory is that individuals with higher potassium intake may also have healthier diets and lifestyles in general, since fruits and vegetables are the foods most likely to be potassium-rich. In other words, the correlation between higher potassium and lower blood pressure may be spurious. However, clinical trials modifying only potassium intake have confirmed the connection.

A recent meta-analysis of randomized controlled trials found that potassium supplementation was able to produce modest but significant declines in blood pressure of 4 to 6 points systolic and 2.5 to 4 points diastolic among patients with essential hypertension (high blood pressure that doesn't have a known cause, also known as primary hypertension).[189] Such reductions are virtually identical to those reported by sodium restriction trials, as a 2003 meta-analysis found. When comparing randomized controlled trials involving either salt restriction or potassium supplementation, Johanna Geleijnse and her colleagues at Wageningen University in the Netherlands found that increasing potassium intake by about 1,700 mg per day was virtually as effective in reducing blood pressure as cutting salt intake by 1,700 mg per day.[190] Increasing potassium by this amount, the equivalent of three to four bananas, may be easier for many people than severely reducing sodium intake.

They estimated that around 98 percent of the U.S. population fails to consume the recommended level of potassium.[191] Low potassium levels may play a role in the development of hypertension and also be related to other diseases, such as diabetes and metabolic syndrome.[192] And, unlike restricting sodium, research indicates that raising the level of potassium in the diet does not adversely affect renal function or blood lipid profiles.[193]

While the dietary guidelines advise increasing the consumption of potassium-rich foods like fruits and vegetables, they ignore the fact that it is practically impossible to get the recommended minimum of potassium *and* stay under the sodium maximum.[194] As a result, the public is unaware that they are being asked to make a choice between a diet low in salt and a diet high in potassium. With salt receiving the lion's share of attention over the past 40 years, people may be unaware that increasing their consumption of potassium *instead of* reducing sodium in their diet might actually be a more effective means of reducing their risk of developing hypertension.

Weight Loss and Exercise

Potassium is not the only approach to hypertension reduction that has largely been ignored by the anti-salt campaigners. Researchers since Dahl have known that body mass is a major predictor of the disease.[195] Numerous trials have found that modest weight loss in overweight people produces blood pressure reductions in hypertensives—larger decreases, in fact, than sodium restriction (though the effects are magnified when they are combined).[196] Weight seems to be a particularly important risk factor for people of African heritage, who are more likely to be obese and/or have higher blood pressure than non-Hispanic whites.[197]

Even small increases in physical activity, without weight loss, have been shown to produce significant decreases in blood pressure in both normotensive and hypertensive individuals. For example, one small trial in 1990 found that hypertensives who engaged in three 40-minute sessions of exercise a week were able to reduce mean arterial pressure as much as they would have on anti-hypertensive drugs and 10 times more than with sodium restriction alone.[198]

Although few of the nation's health organizations have given much attention to these other approaches to hypertension reduction, the general population appears to be growing more skeptical of the sodium reduction guidelines. And predictably, the advocacy organizations that initially went after salt have turned their ire on a new target: sugar.[199] This time around those in charge of providing dietary recommendations will do well to recognize that sugar, like any single nutrient, is just one part of the diet and one factor among many that influence health and that no one factor is the cause of or the solution to health problems for all.

Conclusion

There is little doubt that medical professionals who advocate universal salt reduction are motivated by their desire to improve public health. But their efforts to preserve the demonstrably erroneous salt-hypertension hypothesis violate a core tenet of medical ethics: first, do no harm. The existing evidence indicates that the only group that might benefit from salt restriction are salt-sensitive hypertensives. For the rest, however, reducing sodium appears to be difficult (if not impossible), has no health benefits, and poses significant health risks. As such, it is long past time for health authorities to reconsider their approach to hypertension risk reduction and end the campaign against salt.

The truth is that the science on salt is still in the early phase and that our understanding of its effect on health is limited. What is becoming clear, however, is that the effect of salt intake on health, as with most nutrients, depends on individuals and their context. Some groups, like salt-sensitive hypertensives, might benefit from lowering salt intake, but for most the harms associated with salt reduction outweigh its potential benefits. This emerging evidence is neither perfect nor definitive, but it should give health authorities pause about making blanket recommendations on salt or hypertension-risk reduction.

The best strategy to prevent or treat hypertension, and most other health risks, is one that is individualized and holistic; it must account for individual risk factors, family history, and personal preferences. For some, this approach might very well include salt restriction in addition to other dietary modifications. But the continued messaging from health authorities that salt restriction is the *only* approach has done a great disservice to the public by obscuring other strategies shown to be more beneficial for health. With our currently limited understanding of what drives our salt intake, the benefits high-sodium diets may have, and what effects low-salt diets might have on individual and population health, our current policy of pushing sodium reduction to the entire population is ineffective.

Given the evidence that does exist, government agencies and health organizations—if they are going to do anything—should refocus attention away from a salt-only approach and instead emphasize dietary strategies that promise to result in net benefits for a larger portion of the population and are unlikely to cause negative unintended consequences. Advising people to consume much more of their daily calories from fruits and vegetables, which would increase potassium and other nutrient intakes, based on current research, would likely result in lower blood pressure for a larger portion of individuals without having negative unintended consequences. An approach focused on increasing water and potassium intake by eating more fruits and vegetables may be easier for consumers to follow than significantly reducing sodium and promises to have health benefits beyond and in addition to blood pressure reduction.[200]

How did we get here? The hallmarks of the salt-hypertension issue are familiar. Well-meaning people in public health convince themselves that their position is a moral one. They and their colleagues participate in mutual virtue-signaling, which leads to denigration and ad hominem attacks on those who disagree. Powerful politicians advocate the "virtuous" course and likewise subject skeptics to ad hominem attacks, which further discourages public skepticism. The same politicians dangle the prospect of research funding in front of those who anticipate significant professional advancement if such funding is forthcoming. As a result, published science that increasingly supports previously stated (and funded) hypotheses becomes biased in the direction of funding. A "consensus" is announced by science campaigners like Al Gore and Henry Waxman, and anyone who departs from it is suspected of doing so for nefarious reasons. This, unfortunately, is the cosmos of so much of modern science when regulations, politics, and public funds are involved.

DEATH
The Unintended Consequence of the War on Opioids

T he United States and much of the developed West is beset by an "opioid crisis." Thousands of people each year are victims of opioid overdoses, almost always the result of nonmedical use. Despite concerted international, national, and local efforts to curtail the availability and consumption of opioids, the underground market in opioids continues to thrive, and overdose deaths continue to mount. Societal attitudes toward opium and its derivatives drive opioid policy. These attitudes have fluctuated over the millennia, but in most of the modern era they have been marked by a lack of understanding about the true nature of these drugs, resulting in public policy driven by fear.

For public policy to succeed, it must be evidence-based, unshackled from moral and historical prejudice. Pharmacological, psychological, and medical considerations are crucial. Historical aspects are equally important—for history provides context.

Opium has been used and abused since antiquity. Homer described its preparation in his writings, and its name is believed to derive from the Greek word for "juice," *opos*, referring to the liquid extracted from the unripe seed capsule of the opium poppy. Most historians and archeologists agree that the Sumerians, in what is modern-day Iraq, were the first to cultivate the poppies (*Papaver* somniferum) and isolate opium from their seed capsules, around the end of the third millennium BC. The Sumerians called the poppy *hul gil*, meaning "plant of joy." Originally used to induce euphoria in religious rituals, its medicinal uses, primarily to relieve pain and induce sleep (narcosis), were described as far back as 1500 BC. In the eighth century AD, Arab traders brought the plant to India and China, and by the 13th century AD, it had made its way through Asia Minor to all parts of Europe. Most often the drug was smoked. With the drug came the problem of addiction. Manuscripts have been found from Turkey, Germany, England, and Egypt, dating as far back as the 16th century, that describe the use and abuse of the drug.[1]

In 1806, the German pharmacist Friedrich Serturner isolated the active ingredient in opium and named it *morphine*, after the Greek god of dreams Morpheus. Codeine was isolated from the plant a few years later. After the invention of the hypodermic syringe and hollow bore needle in the 1850s, morphine began to be used intravenously for the treatment of pre- and postoperative pain as well as an anesthetic adjunct.

Diacetylmorphine, or diamorphine, was first synthesized in the laboratory by the English chemist C. R. Alder Wright in 1874. It became popular only after it was independently resynthesized by the chemist Felix Hoffmann, working for the Bayer pharmaceutical company in Germany in 1895. Bayer originally marketed it over the counter as what was thought to be a nonaddictive substitute for morphine, primarily for use as a cough suppressant. Its potency is two and a half times that of morphine. For this reason Bayer gave it the brand name Heroin, from the German word *heroisch*, meaning "strong, heroic."

Of the opioids commonly in use today, methadone (Dolophine) also has two and a half times the potency of morphine. Hydromorphone

(Dilaudid) is twice as potent as heroin and 6.6 times as potent as morphine. Fentanyl (Sublimaze, Duragesic) has 50 times the potency of morphine. Among the commonly prescribed oral opioids, hydrocodone (Vicodin) is roughly equivalent in potency to oral morphine, and oxycodone is roughly twice as potent. Oral codeine has about one-sixth the potency of oral morphine.[2]

The term "opiate" is used to describe alkaloids derived directly from the opium plant, such as morphine. Opioid refers to semisynthetic and synthetic opiates, such as oxycodone (invented in Germany in 1917), hydrocodone (invented in Germany in 1920), and fentanyl (invented in Belgium in the 1960s), as well as antagonist drugs such as naloxone, and endogenous compounds such as endorphins.

Opioids bind with opioid receptor sites in the central nervous system to produce the effects of euphoria, analgesia, narcosis, and respiratory depression.[3] Seventeen known receptors have been reported, but the principal ones are the *mu*, *kappa*, and *delta* receptors. Research on these principal receptors has revealed them to have subtypes. There are three subtypes of the *mu* receptor: *mu1*, *mu2*, and *mu3*. Subtypes 1 and 3 produce analgesia, euphoria, vasodilation, and physical dependence. Subtype 2 produces respiratory depression and physical dependence, but no analgesia. The *mu2* receptor is the only receptor that produces respiratory depression. All of the known receptors produce physical dependence—that is, chronic use requires larger doses to produce the desired effect, and abrupt withdrawal causes unpleasant symptoms. Opioid tolerance is associated with decreases in receptor sensitivity. The *mu1* receptor develops tolerance more rapidly than the *mu2* receptor. This likely explains why some physically dependent opioid users resort to higher and higher doses to achieve the desired effect and gradually succumb to asphyxiation from respiratory depression.[4]

Safety

Despite the risks of physical dependency and respiratory depression, opioids are remarkably safe. The median lethal oral dose (LD50) of hydrocodone, for example, is 375 milligrams per kilogram (mg/kg) in a rat (30,000 mg for an 80 kg adult, or 4,000 7.5 mg pills).[5] The LD50 for oxycodone is 100 mg/kg in a rat (8,000 mg for an 80 kg adult, or 800 10 mg pills).[6] Unlike for many alternative pain medicines, the long-term use of opioids has not been associated with any significant organic damage.[7] There has been no conclusive evidence that chronic opioid use can cause dementia or cognitive decline.[8] Although long-term use can cause constipation and has been shown to reduce testosterone levels,[9] these conditions are treatable. Thus, many patients are on methadone maintenance indefinitely, even for life, and function productively.

Many healthcare practitioners begin the treatment of pain with nonopioids, such as acetaminophen, COX-2 inhibitors, and other nonsteroidal anti-inflammatory agents (NSAIDs). However, experience has shown that these drugs are less effective for severe and chronic pain. They are also not without risks themselves.[10] Acetaminophen can cause hepatotoxicity and cirrhosis. Chronic use of NSAIDs can damage the gastrointestinal tract and cause gastrointestinal bleeding and ulcers, nephrotoxicity and kidney failure, hypertension, and increased risk for heart attack and stroke.[11]

In 2014, the Centers for Disease Control and Prevention (CDC) reported 9.5 deaths per 100,000 population from essential hypertension and hypertensive renal disease, which are just two potential consequences of chronic NSAID use.[12] In that same year, the CDC reported the number of deaths from opioid overdose at 7.9 per 100,000 population.[13]

NSAIDs are not the only known cause of hypertension and its sequelae. Acute and chronic pain, if untreated, cause hypertension and predispose a patient to cardiovascular disease, among other serious problems.[14] Pain affects the blood pressure and pulse, mediated through the sympathetic nervous system, as well as through a neurohumoral pathway, wherein the pituitary and hypothalamus release adrenocorticotropic hormone (ACTH), which in turn stimulates the adrenal glands to release epinephrine and cortisol, both of which increase blood pressure and pulse.[15]

Inadequately treated pain,[16] along with NSAIDs used to treat pain, thus pose greater threats to public health than do properly used opioids.

The LD50 of the common NSAID ibuprofen is 636 mg/kg in the rat.[17] The LD50 for naproxen is 248 mg/kg in the rat.[18] Acute NSAID overdose may feature convulsions, metabolic acidosis, apnea, coma, and acute renal failure.[19]

Overdoses from opioids are almost always associated with the simultaneous ingestion of potentiating drugs, such as alcohol, hypnotics, benzodiazepines, or other opioids such as fentanyl.[20] Data from the New York City Department of Health show that in 2016, 97 percent of opioid-overdose deaths resulted from mixing opioids with other drugs.[21] The Centers for Disease Control and Prevention cite a study of noncancer chronic pain patients between the ages of 15 and 64 who were maintained on opioids and followed for up to 13 years. One of 550 patients died of an opioid overdose during that period, an overdose rate of less than 0.2 percent.[22] A 2011 review of 154,000 Veterans Health Administration patients who received opioids for pain from 2004 to 2008 found an overdose rate of 0.04 percent.[23] A prospective cohort study of 2.2 million patients throughout the state of North Carolina who were prescribed opioids for pain in 2010 revealed an overdose rate of 0.022 percent, with 61 percent of those overdoses involving the mixing of opioids with other drugs.[24]

It is important to distinguish between physical dependence, which can be overcome with detoxification, and addiction, which is a behavioral disorder that has a biological basis.[25] Investigators as well as policymakers have a tendency to mistakenly use the two terms interchangeably.[26] Perhaps as a result, the risk of addiction to opioids has been greatly exaggerated. In 2012, the National Survey on Drug Use and Health reported one case of abuse or dependence for every 130 opioid prescriptions.[27] A 2016 study from Castlight Health, using a liberal definition of "abuse," estimated a 4.5 percent abuse rate among recipients of opioid prescriptions.[28] A 2010 Cochrane review of noncancer patients on long-term opioids for chronic pain found an addiction rate of less than 1 percent.[29] And a 2013 Cochrane review conducted by researchers in Rome, Italy, also found the risk of addiction below 1 percent and concluded, "The available evidence suggests that opioid analgesics for chronic pain conditions are not associated with a major risk for developing dependence."[30] In contrast, a less rigorous study in 2015 from the University of New Mexico places the rate between 8 and 12 percent.[31] A January 2018 study from researchers at Harvard and

Johns Hopkins Universities reviewed over 568,000 patients in the Aetna Health Insurance database who received prescription opioids for acute postoperative pain between 2008 and 2016 and noted a total misuse rate (using the range of misuse diagnostic codes) of 0.6 percent.[32]

Despite the hyperbolic press coverage surrounding the rising national opioid overdose rate, the evidence shows that opioids are a safe and effective choice for severe and chronic pain, carrying little risk of physical harm, and a low risk for abuse, dependence, and overdose.

The Vicissitudes of Physicians' Prescribing Attitudes

In 1973, Richard M. Marks and Edward J. Sachar wrote in the *Annals of Internal Medicine* that physicians' misconceptions about the effective dose range and the risks of addiction "probably lead to undertreatment with narcotic analgesics, causing much needless suffering in medical inpatients."[33] The authors estimated less than 1 percent of patients treated for pain with narcotics became addicts. They concluded, "For many physicians these drugs may have a special emotional significance that interferes with their rational use."

Today's policymakers often argue that the call, in the 1990s, for a more liberal use of opioids to treat pain was based on nothing more than a 1980 letter to the editor of the *New England Journal of Medicine* by Jane Porter and Hershel Jick,[34] which emphasized the low risk for addiction attending opioid use in hospitalized patients. But the letter cited comprehensive drug surveillance performed by their collaborative program at Boston University.[35] These findings were supported by subsequent studies, published by various researchers in the 1980s, that found the risk of addiction of hospitalized patients as well as chronic pain patients treated with narcotics was negligible.[36] (As previously mentioned, numerous contemporary studies, including 2010 and 2013 Cochrane analyses, continue to yield these findings.)[37] Nevertheless, doctors persisted in their fear of prescribing opioids to treat their patients in pain.

In a 1987 article in the *New York Times* headlined "Physicians Said to Persist in Undertreating Pain and Ignoring the Evidence," Dr. Russell Portnoy, director of analgesic studies in the Pain Center at Sloan-Kettering Memorial Hospital, said, "The undertreatment of pain in hospitals is absolutely medieval."[38] The War on Drugs subjected medical students and physicians to the same opiophobia it subjected the general public to. In the *Times* article, Portnoy stated, "The problem persists because physicians share the same widespread social attitudes that these drugs are unacceptable."

In 1989, the National Institute on Drug Abuse (NIDA) acknowledged this problem. NIDA director Charles Schuster stated, "We have endowed these drugs with a mysterious power to enslave that is overrated." A 1993 NIDA newsletter stated, "These drugs are rarely abused for medical purposes," and, "thousands of patients suffer needlessly."[39] In 1992 and 1994, the U.S. Department of Health and Human Services issued guidelines encouraging the more aggressive management of patients in pain.

By the mid-1990s, the American Pain Society exhorted healthcare providers to consider pain as the "fifth vital sign," after the first four of body temperature, pulse rate, respiration rate, and blood pressure.[40] Soon thereafter, the Joint Commission on Accreditation of Healthcare Organizations adopted the same policy and promoted continuing education seminars on the subject. A book published by the Joint Commission in 2000 stated, "There is no evidence that addiction is a significant issue when persons are given opioids for pain control."[41]

Healthcare practitioners were not the only obstacle to a more rational approach to pain control. Patients also suffered from opiophobia. Many feared that even a small amount of narcotic for their pain would lead to instant addiction. Over time, as healthcare personnel began treating pain as the fifth vital sign, patient attitudes toward opioids changed.

The CDC reports that opioid prescriptions tripled from 1999 to 2015.[42] Opioid overdose deaths quadrupled during the same timeframe.[43] A recent review of data from the National Survey on Drug Use and Health and the CDC show that no correlation exists between the number of opioid prescriptions and in the incidence of "past month nonmedical use of pain relievers" in adults and "pain reliever use disorder in the past year" among adults.[44]

An increase in the supply of opioids would be expected to lead to an increase in opioids available for diversion to nonmedical users.

But the forces behind nonmedical use are multiple, and involve socioeconomic and psychosocial, as well as sociocultural factors.[45] A November 2017 study by T. J. Cicero and colleagues found that 33 percent of heroin users who started treatment in 2015 commenced opioid abuse with heroin (as opposed to 8.7 percent who commenced with heroin in 2005), demonstrating that an increasing number of addicts do not get their start with prescription pills.[46] A 2018 study from the University of Pittsburgh Medical Center reviewed overdose deaths from nonmedical use of illicit drugs using data from the CDC dating back to the 1970s. The researchers concluded, "The U.S. drug overdose epidemic has been inexorably tracking along an exponential growth curve since at least 1979. . . . This historical pattern of predictable growth for at least 38 years suggests that the current opioid epidemic may be a more recent manifestation of an ongoing longer-term process."[47] An NBC News report summarized the study findings this way: "Substance abuse stared before the availability of synthetic opioids and may have only a little to do with prescribing habits of doctors or the pushy habits of drug makers."[48]

However, many policymakers have now concluded, despite years of research on the subject, that the direct cause of the opioid overdose rate is the liberalized prescription of opioids by healthcare providers. This is a return to the pre-1990s belief, concretized in the Eugene O'Neill play *A Long Day's Journey into Night*, that doctors prescribe opioids to their patients in pain who, in turn, fall prey to the addictive powers of the drug and descend into the dark world of drug abuse. Yet, in the European Union, where the prescription of opioids for pain has historically been more restrictive than in the United States, opioid overdoses continue to rise as well.[49]

Recrudescent Opiophobia

By 2010, state and local public health efforts were instituted to restrict the prescription and distribution of opioids. These efforts were further stimulated by media reports of arrests of doctors operating "pill mills."[50] A "pill mill" is an operation in which a healthcare provider, clinic, or pharmacy prescribes and/or dispenses narcotics without a legitimate medical purpose.[51] In effect, these operators are using their healthcare licenses as a cover to sell opioids for nonmedical use. The existence of these "bad apples" fueled the narrative that healthcare practitioners were the cause of rising overdoses.

Starting in the early part of this century, states instituted prescription drug monitoring programs to maintain surveillance on narcotics prescriptions as well as on those patients receiving them.[52] By 2017, all 50 states plus the District of Columbia had these programs. (Missouri, the last of the 50 states to establish a program, did so only after its governor issued an executive order in the summer of 2017.) Early research indicates that they have had a chilling effect on prescribers' activities, but have done nothing to reduce the overdose rate, and may have driven desperate pain patients to the illicit drug market in search of relief.[53]

The CDC has actively undertaken efforts to curtail opioid prescriptions and continues to update guidelines for prescribers.[54] The prescription drug rate has decreased every year since it peaked in 2010, and prescriptions for higher and more dangerous doses have dropped 41 percent since 2010.[55] The Drug Enforcement Administration ordered a 25 percent reduction in the manufacture of opioids in 2016, and recommended another 20 percent reduction in 2017.[56]

The Food and Drug Administration has encouraged the development of abuse-deterrent formulations of prescription opioids that are unsuitable for snorting or injecting.[57] In 2010, the makers of OxyContin made the abuse-deterrent formulation the only one available on the market. This opportunity for pharmaceutical manufacturers to "evergreen" their patents has encouraged other manufacturers to develop their own abuse-deterrent formulations of other opioids. "Evergreening" is the practice of using existing legal means to extend the life of patents about to expire, in order to retain the royalties and/or market dominance they provide. Reformulating a drug to modify its properties is an approach to "evergreening" in common use. Meanwhile, numerous studies suggest the abuse-deterrent formulations only foster the migration of nonmedical users to heroin.[58] A June 2017 working paper by economists at Notre Dame University and Boston University found a "one-to-one substitution of heroin deaths for opioid deaths" for OxyContin since its replacement with its abuse-deterrent formulation.[59]

Meanwhile, heroin and fentanyl, smuggled in from other countries, have flooded the market and are usually cheaper and more readily available than diverted prescription opioids.[60] In 2015, CDC director Thomas Frieden estimated that heroin sold for one-fifth the street price of prescription opioids.[61] Some dealers on the black market sell counterfeit prescription opioid capsules filled with fentanyl.[62] Heroin is increasingly being laced with fentanyl to enhance its potency, a practice that also increases the risk of overdose.[63]

Unintended Consequences of the New Opiophobia

According to the National Survey on Drug Use and Health (NSDUH), nonmedical use of prescription opioids peaked in 2012 and is now below the rate in 2002, and total opioid use, including heroin, was even lower in 2014 than in 2012. The NSDUH also reported in 2014 that only 25 percent of nonmedical opioid users got them by obtaining a doctor's prescription.[64] The overwhelming majority of overdose patients presenting to emergency rooms are not patients receiving prescription opioids for pain.[65] Despite these data, the annual overdose death rate continues to climb. In 2015, the number of deaths climbed to approximately 33,000, but for the first time since the CDC began collecting this data, the majority of these deaths were due to heroin and fentanyl, while the overdose rate from prescription opioids was found to have nearly stabilized.[66] The CDC overdose report for 2016 was released in December 2017.[67] Preliminary data pointed to a further increase in overdose mortality, with an even larger heroin and fentanyl component,[68] and this was borne out in the final report. Fentanyl is now emerging as the primary cause of opioid overdose deaths, with heroin close behind in second place.[69]

Restrictive new policies toward opioid prescription, production, and distribution took effect in 2010. According to data provided by the CDC,[70] the five-year trend line from 2006 to 2010 predicted an overdose rate from prescription opioids of 5.75 per 100,000 and an overdose rate from heroin and fentanyl of 2.62 per 100,000 in 2015. The actual

FIGURE 4.1

Overdose deaths involving opioids, by type of opioid, United States, 1999–2016

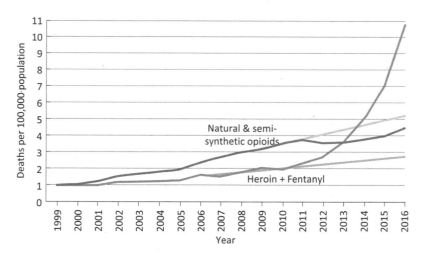

Source: Centers for Disease Control and Prevention, CDC Wonder, "Current Multiple Cause of Death Data," https://wonder.cdc.gov/mcd.html.

overdose rate for prescription opioids was 4.84 per 100,000, and the actual overdose rate for heroin and fentanyl was 6.30 per 100,000. The data suggest that one fewer person per 100,000 is dying from prescription opioid overdose after five years of restrictive policies in exchange for nearly four more people per 100,000 dying from heroin and fentanyl (see Figure 4.1).

CDC data on the opioid prescription rate goes only as far back as 2006. The overdose death rate as a proportion of opioid prescriptions from 2006 to 2010 remained stable at roughly one overdose per 13,000 prescriptions. After 2010, there is an inverse relationship between the number of prescriptions written and the overdose rate (see Figure 4.2).

The data suggest that the resumption of opiophobia and its resultant focus on the availability of prescription of opioids to treat pain not only fails to reduce overdose deaths, but makes matters worse. Tightening the supply of prescription opioids available to the illicit market for nonmedical users, in an environment where cheaper and more dangerous heroin and fentanyl are readily available, only

FIGURE 4.2

Overdose deaths involving opioids, by type of opioid, compared to opioid prescription rate, United States, 1999–2015

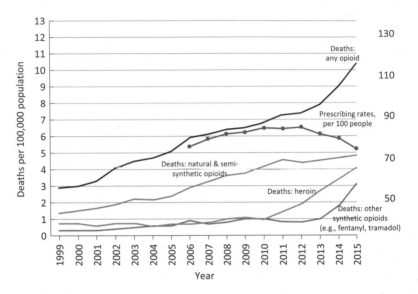

Source: Centers for Disease Control and Prevention, CDC Wonder, "Current Multiple Cause of Death Data," https://wonder.cdc.gov/mcd.html.

drives nonmedical users to those more dangerous alternatives. In the meantime, the curtailment of prescriptions is reigniting the rise in the number of patients who are undertreated for their pain. Some, in desperation, seek relief in the black market, where they are exposed to dangerous, adulterated drugs.[71]

Conclusion: Opiophobia, Rinse, and Repeat

Since antiquity, the opium poppy and its derivatives have been valued for their ability to induce euphoria and narcosis and to relieve pain. Numerous studies, dating back more than 50 years, demonstrate the safety and efficacy of opioids, their low addictive potential in the medical setting, the negligible deleterious organic effects of long-term use, and their low potential for overdose when not used in conjunction with other potentiating drugs.

The propaganda surrounding drug prohibition led to an irrational fear of opioids on the part of healthcare practitioners and patients until the late 1980s. This resulted in an unfortunate undertreatment of patients in pain. As this became more evident to public health officials, providers, and patients, a corrective process led to an increase in the prescription and use of opioids, as well as a more compassionate focus on the treatment of pain. It also led to an improvement in the drugs' reputation among the general public.

Unfortunately, the economic incentives that drug prohibition provides to drug dealers meant that the increase in the production, prescription, and availability of these opioids offered an easy target for those seeking to divert the drugs to the illicit market for recreational use. Furthermore, unscrupulous members of the healthcare professions and allied fields took advantage of the enlightened attitude toward opioids and pain management and gamed the system for personal profit, feeding a growing market of nonmedical users. Meanwhile, a complex mix of socioeconomic, psychosocial, and sociocultural changes affected the nonmedical use of opioids and other narcotics.

During the first decade of the 21st century a dramatic increase in opioid addiction and overdose deaths seized the national spotlight. Failing to recognize that the deaths were primarily a result of drug prohibition—where the dose and purity of any substance obtained is always suspect—and that the causes of abuse and addiction were multifactorial, public health officials and policymakers on the state and national level resumed the opiophobic biases and proclivities of earlier years. Restrictions were reinstituted on the production, distribution, and prescription of opioids for pain. Many patients in pain retrieved their discarded fear of becoming easily addicted to opioids and needlessly deny themselves pain medication. Still, other patients in chronic pain find themselves desperately cut off from the source of their relief.

The unintended consequences of opiophobia in the latter part of the 20th century led to the undertreatment of pain and the avoidable anguish of countless patients. The unintended consequences of opiophobia in the 21st century are even worse. Opiophobia is now not only responsible for needless pain; it is also responsible for needless deaths.

Summary

It is widely believed that these events launched the opioid epidemic:

- Doctors prescribed opioids legitimately for pain relief.
- Patients then became dependent on or addicted to opioids.
- Patients created a black market for opioids long after their pain had resolved.
- That black market has led to over 49,000 deaths a year in the United States.[72]
- Medically prescribed opioids are therefore the "gateway" to the opioid epidemic.

That narrative is false. Actually,

- Medically prescribed opioids are surprisingly safe, in that deaths from overdose are rare. Moreover, medically prescribed opioids precipitate remarkably little dependence or addiction.
- Rather, people who are drug abusers independently of medical prescription abuse opioids by virtue of their availability.
- We should therefore continue to encourage physicians to prescribe opioids to patients whose pain would otherwise go undertreated.

DRUGS

The Systematic Prohibition of U.S. Drug Science

The drug war is not and has never been based in science. Early drug warriors did not commission scientific studies to rank the most medically and socially harmful substances and then prohibit them accordingly. Science rarely played a part in deciding which drugs would be prohibited; in fact, where there was scientific research, drug warriors often denied and opposed it.

It is crucial to remember this fact when examining how governments have influenced scientific research into drugs. Unlike the subjects of other chapters in this volume, restrictions on drug research are the primary method by which the government has tilted drug science. Although those who study diet, energy, and climate science have encountered problems with funding and biased government agencies, they rarely have to deal with the outright prohibition of the objects of their research.

Drug prohibition has developed primarily from a confluence of three factors: (1) the popular perception of the race or class of the drug user, (2) the popular perception of the drug's effects, and (3) the dehumanization of the perceived drug user. Drugs have often been prohibited before any scientist had a real chance to study the effects and before most politicians and voters had interacted with the drug in their daily lives. Rumor, myth, and prejudice explain drug prohibition, not science.

The process that leads to prohibition plays out with an almost eerie regularity—a scripted public-policy play performed in legislatures around the world and throughout the 20th and 21st centuries. First, reports begin to circulate about a disturbing new drug with a strange name like "demon weed," "ecstasy," "acid," or "bath salts." Law enforcement officials often give the first accounts of the drug's dangerous effects, which biases the story because law enforcement officers are not usually dealing with people at their best moments (imagine if the public perception of alcohol were informed only by stories of police interacting with drunks). "It's like nothing we've seen before," police say; "the people on this drug are not only maniacs, but they are also nearly unkillable." Tales of Hulk-like strength seem to follow every drug.

The drug is soon associated with a certain race, class, or nationality—perhaps African Americans, Chinese, Mexicans, or even hippies—and parents are warned that their children may already be taking it. Talking heads with dubious scientific credentials take to print and the airwaves to say that the drug is the most addictive drug they've seen and that even a single use could create a lifelong addict. Soon anyone who uses the drug, and especially those who manufacture and sell it, are dehumanized as subhuman scum, "drug fiends," and "merchants of misery" who must be stopped no matter the cost.

After a drug is restricted or outright prohibited, any scientist who wants to study it must contend with new layers of control and bureaucracy. Those extra restrictions are significant barriers to scientific research—particularly if that research will inquire into the possibility that an illegal drug has beneficial uses. Funding also becomes relatively difficult to obtain, especially from the government. By prohibiting or increasing regulation, governments around the world bias the scientific study of drugs.

Moreover, for many scientists, even to propose studying a prohibited drug in a clinical, scientific manner could be dangerous to their career and professional standing. One politician, when asked how his views on marijuana prohibition might change if scientists came out favorably for marijuana wondered, "Why should we use it when it has no redeeming value? The desire of someone to get out of this world by puffing on marihuana [sic] has no redeeming value."[1] President Richard Nixon, when asked how he would react to the report of a commission that was studying marijuana, was even more direct: "As you know, there is a Commission that is supposed to make recommendations to me about this subject, and in this instance, however, I have such strong views that I will express them. I am against legalizing marijuana. Even if the Commission does recommend that it be legalized, I will not follow that recommendation."[2]

Faced with headwinds like these, it's no wonder that scientists often don't bother researching controlled substances. Results that find beneficial uses of drugs or call into question a drug's dangerousness can be criticized by politicians who are either dogmatic drug warriors or who do not see a political benefit to sticking their necks out to advocate for real change.

That gun-shy behavior of political actors was demonstrated in 2012 during an almost comical congressional committee hearing on Drug Enforcement Administration (DEA) oversight. Michele Leonhart, administrator of the agency, was asked by Rep. Jared Polis (D-CO) about the relative dangers of different drugs:

POLIS: Is crack worse for a person than marijuana?
LEONHART: I believe all the illegal drug —
POLIS: Is methamphetamine worse for somebody's health than marijuana?
LEONHART: I don't think any illegal drug —
POLIS: Is heroin worse for someone's health than marijuana?
LEONHART: Again, all the drugs —
POLIS: I mean, either yes, no, or I don't know. I mean, if you don't know, you can look this up. You should know this as the chief administrator for the Drug Enforcement Agency. I'm asking you a very straightforward question. Is heroin worse for someone's health than marijuana?
LEONHART: All the illegal drugs are bad.
POLIS: Does this mean you don't know?

LEONHART: Heroin causes an addiction that causes many problems that's very hard to kick.

POLIS: Does that mean that the health impact is worse than marijuana, is that what you're telling me?

LEONHART: I think that you are asking a subjective question.

POLIS: No. It is objective. Just looking at the science. This is your expertise. I am a lay person, but I have read some of the studies and [am] aware of it. I am just asking you as an expert in the subject area, is heroin worse for someone's health than marijuana?

LEONHART: I am answering as a police officer and as a DEA agent that these drugs are illegal, because they are dangerous, because they are addictive, because they do hurt a person's health.

. . .

MR. POLIS: Well, again, this is a health-based question, and I know you obviously have a law enforcement background, but I am sure you are also familiar, given your position with the science of the matter, and I am asking, you know, again, clearly, your agency has established abuse of prescription drugs as the top priority. Is that, therefore, an indication that prescription drugs are more addictive than marijuana?

LEONHART: All illegal drugs are addictive.[3]

Despite serving under President Obama, Leonhart was a holdover from the George W. Bush administration, and she had become a DEA agent in late 1980, just before President Ronald Reagan ramped up the drug war.[4] (Leonhart also reportedly once said that the day a hemp U.S. flag flew over the Capitol building was the worst day of her life.)[5] It is perhaps understandable that Leonhart would hold unnuanced views on the relative dangers of the drugs she has dedicated her life to fighting.

But the political position of the DEA administrator demands unnuanced views on drugs. The DEA is tasked with eliminating drugs, not promoting them. It has been the policy of the U.S. government since the early 1970s that marijuana is as dangerous as heroin, LSD, MDMA (ecstasy), and bath salts. Those drugs are placed in Schedule I of the Controlled Substances Act (CSA), meaning that, according to the government, they have a high potential for abuse, there are no currently accepted medical uses, and there is a lack of "accepted safety" standards for use of the drugs.[6] While there are good reasons to believe that many scheduled drugs are misclassified (there are five schedules, with I deemed the most dangerous), Schedule I is truly the Hotel California of government classifications—you can check out any time you like, but you can never leave. In over 40 years, the DEA has only rescheduled a Schedule I drug seven times: five times a Schedule I drug has been moved to Schedule II, and only twice has a Schedule I drug been descheduled entirely.[7]

This chapter discusses how the government has distorted scientific research on two Schedule I drugs, marijuana and MDMA, also known as ecstasy. Although a whole book could be written about how the government has affected drug research for a variety of illicit drugs, I focus on marijuana and MDMA for two reasons. First, government restrictions have had a unique and unmistakable effect on biasing the scientific study of these two drugs, and second, because there is mounting evidence that these drugs are both not particularly dangerous and can provide significant benefits to users suffering from a variety of ailments.

Marijuana

In August 2016, despite recreational marijuana being legal and commercially available in four states, and despite 25 states and the District of Columbia allowing at least medicinal marijuana, the DEA reviewed its scheduling of marijuana and decided that it would remain a Schedule I drug.[8] It was the fourth time since 1972 that the DEA had denied a petition calling for the rescheduling of marijuana. The fate of those four petitions encapsulates the government's institutional recalcitrance on neutral drug research and the difficulties faced by anyone who wants to change our outdated laws.

The CSA, signed into law by President Nixon in 1970, was supposed to make reconsideration of drug scheduling relatively straightforward, at least as straightforward as dealing with the government can be. The law initially listed 81 substances to be placed on Schedule I, including marijuana. But in a concession to some concerned lawmakers, it also created a commission to study marijuana use.[9] Drug classifications under the CSA can be reviewed at the behest of the attorney general, the secretary of health and human services, or an interested third party.[10]

Scientific and medical evidence were to be the basis for CSA classification. In fact, during the drafting of the act, an amendment was proposed that would have expressly required scientific evidence to form the basis for scheduling. But the amendment was defeated as being redundant. In the words of Bureau of Narcotics and Dangerous Drugs (the precursor to the DEA) director John Ingersoll:

> The bill allows the Attorney General upon his own motion or on the petition of an interested person to bring a drug under control. However, he is authorized to do so only after requesting the advice in writing of the Secretary of Health, Education, and Welfare and the advice in writing of the Scientific Advisory Committee. . . . The intent of the amendment was to insure that the scientific and medical information necessary for a determination of whether a substance should be brought under control was available. But the legislation already insured that there would be sufficient medical and scientific input into any control decision.[11]

The first third-party petition for marijuana rescheduling was filed in 1972 by the National Organization for the Reform of Marijuana Laws (NORML). The DEA was created the following year, and it promptly refused to consider NORML's petition. NORML went to court, and in 1974 the United States Court of Appeals for the District of Columbia Circuit ordered the DEA to comply with the statute and review the petition.[12]

In response to the court order, the DEA held a three-day hearing before an administrative law judge. The judge agreed with NORML on many counts, yet the DEA administrator entered a final order denying the petition.[13] NORML went back to court and won again, with the court ordering the DEA to submit the petition to the secretary of health, education, and welfare (HEW).[14] The HEW secretary recommended that marijuana remain on Schedule I, and the DEA ratified that determination without further hearings. The DEA's cursory actions led NORML to bring yet another challenge to the courts. Again the court not only agreed with NORML, but found it necessary to chide the government for continually ignoring its orders: "[We] regrettably find it necessary to remind respondents [DEA and HEW] of an agency's obligation on remand not to 'do anything which is contrary to either the letter or spirit of the mandate construed in the light of the opinion of [the] court deciding the case.'"[15]

It wasn't until 1986, 14 years after the original petition, that the DEA agreed to hold substantive hearings on rescheduling that complied with the requirements of the CSA. From October 1987 until June 1988, administrative law judge Francis L. Young heard evidence on whether marijuana's medical applications were sufficient to demand a "demotion" to Schedule II. In his ruling, Judge Young found that marijuana should be rescheduled:

> Based upon the foregoing facts and reasoning, the administrative law judge concludes that the provisions of the Act permit and require the transfer of marijuana from Schedule I to Schedule II. The Judge realizes that strong emotions are aroused on both sides of any discussion concerning the use of marijuana. Nonetheless it is essential for this Agency, and its Administrator, calmly and dispassionately to review the evidence of record, correctly apply the law, and act accordingly.

> Marijuana can be harmful. Marijuana is abused. But the same is true of dozens of drugs or substances which are listed in Schedule II so that they can be employed in treatment by physicians in proper cases, despite their abuse potential.[16]

Amazingly, the DEA still refused to follow Judge Young's recommendation. In December 1989, DEA administrator John Lawn overruled Young and in 1994, 22 years after NORML first filed the petition, the DC Circuit upheld Lawn's decision.

NORML's ridiculous journey through the red tape of the DEA underscores the political and legal headwinds that have restricted and are restricting the implementation of a drug policy based in neutral scientific and medical research. Behind the DEA's determined foot-dragging is the dogged resistance that arises when an organization tasked with eradicating drugs is asked to make one of those drugs more accessible. This attitude toward drugs—that elimination through prohibition is the only viable solution—permeates many of the organizations that oversee both funding research and allowing researchers to access controlled substances.

What's more, NORML's petition was not the last time the DEA denied petitions to review the scheduling of marijuana. Three other times the DEA has refused to accept that marijuana has enough medical uses to be considered at most a Schedule II drug.

Unsurprisingly, the deck is stacked against marijuana when it comes to petitioning for rescheduling. Because of its Schedule I status, it is difficult for researchers to study marijuana to produce the kind of clinical evidence that could help marijuana be rescheduled. Thus, while the DEA requires "proof" of marijuana's medical effectiveness, government policies are preventing such studies from being done.

During one chapter in NORML's 22-year fight with the DEA, the DC Circuit highlighted this unfairness. When the DEA overruled Judge Young's findings, it did so partially on the grounds that marijuana had failed to meet three criteria: (1) the general availability of the substance and information regarding the substance and its use; (2) recognition of its clinical use in generally accepted pharmacopeia, medical references, journals, or textbooks; (3) recognition and use of the substance by a substantial segment of the medical practitioners in the United States.[17] NORML argued, and the court agreed, that it was impossible for any drug classified as Schedule I to meet these criteria. As the court wrote:

One of the very purposes in placing a drug in Schedule I is to raise significant barriers to prevent doctors from obtaining the drugs too easily. DEA regulations require doctors who wish to use such drugs to submit a scientific research protocol to the FDA for approval and permit use only in accordance with the protocol. And the FDA insists that a developed scientific study program be presented in order to gain approval of the protocol. The DEA regulations further impose mandatory registration with the DEA and mandatory record-keeping and safe-keeping requirements, presenting additional barriers to widespread use. We are therefore hard-pressed to understand how one could show that any Schedule I drug was in general use or generally available. We are also concerned that the fifth factor "recognition of [a drug's] clinical use in generally accepted pharmacopeia, medical references, journals, or textbooks" might be subject to the same objection. Petitioners assert that if a drug is not widely prescribed—regardless of its safety or use—it will not appear in a pharmaceutical listing of medically useful drugs.[18]

The DEA eventually figured out a way around this legal snafu by subtly changing the criteria used. Yet that did not resolve the fundamental issue of government interference with marijuana research, a problem that has bedeviled marijuana researchers for decades.

Although the barriers to scientifically studying marijuana have improved in modern times, there are still significant problems. In the United States, as well as around the world, various United Nations treaties restrict drug research, including the 1961 Single Convention on Narcotic Drugs, the 1971 Convention on Psychotropic Substances, and the 1988 Convention against Illicit Traffic in Narcotic Drugs and Psychotropic Substances. The 1971 treaty requires parties to "prohibit all use except for scientific and very limited medical purposes by duly authorized persons, in medical or scientific establishments which are directly under the control of their Governments or specifically approved by them."[19]

The United States takes its obligations under these treaties very seriously, especially when it comes to studying marijuana. In fact, despite the fact that marijuana is by far the most commonly used illicit substance in the world and its medical applications are now recognized by 32 states and the District of Columbia, the federal government makes studying marijuana more difficult than studying other Schedule I drugs. While it is comparatively easy for researchers to get permission from the DEA to produce and study Schedule I compounds like LSD and MDMA (although it is still very difficult), marijuana researchers must go through three different government bureaucracies. First, a license must be granted by the DEA. Second, to do clinical research, further approval by the Food Drug Administration (FDA) is required. Finally, to actually obtain marijuana, researchers must go through the National Institute on Drug Abuse (NIDA).[20]

Yet even reviewing the steps in that cumbersome process does not adequately convey the barriers to scientifically neutral cannabis research. NIDA, after all, is an organization focused on drug abuse, not medicinal, therapeutic, or other positive uses. According to Dr. Steven Gust, former special assistant to the director of NIDA, NIDA's mission is to "support research on the causes, consequences, prevention, and treatment of drug abuse and addiction."[21] It is not NIDA's mission, according to Dr. Gust, "to study medicinal uses of marijuana or to advocate for such research."[22]

NIDA controls the only source through which researchers can legally acquire marijuana. That source is the University of Mississippi, which for nearly 50 years has had a monopoly on sanctioned marijuana production. Nothing in U.S. law mandates only a single source of marijuana production, and in fact the Controlled Substances Act requires an "adequate and uninterrupted supply of these substances under adequately competitive conditions for legitimate medical, scientific, research, and industrial purposes."[23] Nevertheless, although many research organizations have applied for permission to also grow marijuana, all have been denied.

The DEA has said that "for most of the nearly 50 years that this single marijuana grower arrangement has been in existence, the demand for research-grade marijuana in the United States was relatively limited—and the single grower was able to meet such limited demand."[24] Yet in 2007, the DEA's own administrative judge found that "NIDA's system for evaluating requests for marijuana research has resulted in some researchers who hold DEA registrations and the requisite approval from the Department of Health and Human Services being unable to conduct their research because NIDA has refused to provide them with marijuana."[25] For example, in 2011 the Multidisciplinary Association for Psychedelic Studies (MAPS) sought cannabis for an FDA-approved study of post-traumatic stress disorder (PTSD), but NIDA denied access because approval had not been given by the Public Health Service (this requirement was eliminated in 2015). After receiving Public Health Service approval, MAPS was told that NIDA did not have the cannabis required for the study. It wasn't until April 2016 that the study finally moved forward.[26] These types of delays, if not outright obstruction, have been quite common, which may explain why the DEA perceived its supply as adequate to meet "such limited demand"—researchers simply didn't bother wading into the DEA's bureaucratic morass and chose other things to research instead.

MAPS has been continually stymied by the lack of an adequate supply of NIDA-produced marijuana. Although NIDA eventually provided marijuana for MAPS to study the treatment of PTSD in veterans (as of March 2017, this study had moved into phase 2 of clinical trials),[27] the government has simply blocked other attempts to study marijuana. Since June 2003, for example, NIDA has blocked research into whether vaporizing marijuana can effectively mitigate some of the harms cause by smoking the drug through combustion. According to MAPS:

In a prolonged triumph of drug-war politics over science, our vaporizer research has been blocked since June 2003 by the National Institute on Drug Abuse (NIDA), which has a monopoly on the supply of marijuana that can be used in research. Since 2003, NIDA has rejected and/or ignored our repeated requests (including one lawsuit for "unreasonable delay") seeking to purchase 10 grams of marijuana to continue our studies. NIDA uses its monopoly to obstruct studies into both the beneficial medical uses of marijuana as well as into drug delivery devices that might increase the chances of FDA approval of marijuana as a prescription medicine, and might decrease the harms associated with the non-medical uses of marijuana.[28]

In August 2016, under the Obama administration, the DEA switched course and announced it would take applications to authorize other sources of research marijuana. Yet the effects of 50 years of government stifling marijuana research will be tough to overcome. Requirements for new applicants are stiff, and it is still easy for the DEA to simply claim that no applicant meets the requirements. Or the DEA can simply ignore the applications, which seems to be exactly what happened under the Trump administration and former attorney general Jeff Sessions, a committed drug warrior who even asked Congress for the ability to prosecute medical marijuana distributors.[29]

In August 2017, the *Washington Post* reported that "Attorney General Jeff Sessions has effectively blocked the Drug Enforcement Administration from taking action on more than two dozen requests to grow marijuana to use in research."[30] In the year after the Obama administration began accepting applications, 25 proposals were submitted. According to one law enforcement official, "They're sitting on it. They just will not act on these things." "[T]he Justice Department has effectively shut down this program to increase research registrations," said a DEA official.[31] Sens. Kamala Harris (D-CA) and Orrin Hatch (R-UT) twice sent letters to Attorney General Sessions, in April and August of 2018, arguing that it is "imperative that our nation's brightest scientists have access to diverse types of federally-approved, research-grade marijuana to research both its adverse and therapeutic effects" and asking him to at least "put a date on when the DEA will take action on the more than two dozen pending applications."[32] As of May 2019, no applications had been approved.[33]

The ease with which policies can shift between administrations underscores the need for a more permanent legislative fix. Senator Hatch and Sen. Brian Schatz (D-HI) introduced the MEDS Act in September 2017, which would "remove the administrative barriers preventing legitimate research into medical marijuana" by, among other things, making marijuana more available for legitimate research.[34] That is a heartening development, but the bill's future is uncertain. And in September 2018, the House Judiciary Committee approved the Medical Cannabis Research Act, which would require the attorney general to approve at least two qualified suppliers of research-grade marijuana within a year after the act is passed, and at least three qualified suppliers every year after that.[35] As of this writing, the bill will be introduced in the House of Representative, but its future is also uncertain.

If either the MEDS Act or the Medical Cannabis Research Act fails to pass, and even if William Barr, Sessions's replacement as attorney general, allows the application process to move forward, the DEA may have placed a poison pill in the requirements for applicants to grow marijuana for research. Many applicants may not qualify because one "factor" the DEA will consider is "whether the applicant has previous experience handling controlled substances in a lawful manner and whether the applicant has engaged in illegal activity involving controlled substances."[36] This means that experienced growers who have been supplying the state-legal markets in Colorado, Washington, and elsewhere will have a difficult time being approved because "illegal activity includes any activity in violation of the CSA (regardless of whether such activity is permissible under State law)."[37] This bizarre requirement may stifle new applicants altogether. Owing to the costs of complying with the DEA's other requirements, such as controls to prevent diversion of the crop into the general marketplace, few funders may want to take risks on producers with no previous experience growing marijuana. "Why would anyone take a gamble in trying to meet all of these requirements had they not had prior experience with growing!?," asked Dr. Sue Sisley, who conducted a DEA-approved study on treating PTSD with cannabis.[38]

The single-source system is and has been a severe constraint on scientifically neutral research into marijuana, but Dr. Sisley's comment points to another: funding. In fact, in the DEA's comments pertaining to new applications for official marijuana sources, the agency acknowledged in a footnote that funding is often more important than sourcing marijuana. That footnote is worth quoting in full:

Funding may actually be the most important factor in whether research with marijuana (or any other experimental drug) takes place. What appears to have been the greatest spike in marijuana research in the United States occurred shortly after the State of California enacted legislation in 1999 to fund such research. Specifically, in 1999, California enacted a law that established the "California Marijuana Research Program" to develop and conduct studies on the potential medical utility of marijuana. The state legislature appropriated a total of $9 million for the marijuana research studies. Over the next five years, DEA received applications for registration in connection with at least 17 State-sponsored pre-clinical or clinical studies of marijuana (all of which DEA granted). However, it appears that once the State stopped funding the research, the studies ended.[39]

Ironically, those 17 approved California state-funded studies were cited in a 2004 order that denied a MAPS petition to qualify as an official marijuana producer.[40] The fact that NIDA's marijuana was sufficient to supply those studies was cited as evidence that NIDA's supply was adequate and thus no further producers needed to be authorized.

Funding for neutral marijuana research has been difficult to come by, particularly if that funding is sought from the federal government. Some marijuana-friendly states have funded studies, such as the 17 funded by California and Dr. Sisley's PTSD study, which was funded by a $2 million grant from the state of Colorado. Yet researchers who seek funding from a federally controlled entity are likely to come up wanting. According to one study, "$1.1 billion of the $1.4 billion that the National Institutes of Health spent on marijuana research from 2008 to 2014 went to study abuse and addiction," leaving only $297 million to spend on studies looking into "effects on the brain and potential medical benefits for those suffering from conditions like chronic pain."[41] A search of medical studies in 2013 found that only 6 percent were investigating the potential benefits of cannabis rather than its harms.[42]

These statistics confirm the observations of David Nutt, a neuro-pharmacologist in the United Kingdom. As he wrote on his blog:

> Scientists examining the health and social impacts of drugs are usually funded by government agencies, and they often highlight the negative effects of drugs to justify their own source of funding. If a scientist can show that a drug is harmful, then they can show that it's important to do more research on the topic to protect society. The more harmful the drug appears to be, the more

critical it is to fund research on it, so their funding is perpetuated. By contrast, if use of a drug appears to have only benign effects, then why would the government bother spending more research money on it? This cycle has perpetuated a scientific industry intent on demonstrating the harmful effects of drugs like cannabis, precisely because doing so allows the industry to justify its own existence. Of course, one of the major consequences of this cycle is that research on the potential benefits of drugs gets sidelined, for example with not nearly enough government funding for studies on cannabis' health benefits.[43]

It is clear that marijuana's status as a Schedule I drug is a severe hindrance to good scientific research. Our policy choices of the 1960s and 1970s are not serving us well in a modern world that is slowly, but inexorably, coming to terms with the benefits of cannabis, both recreationally and medically.

MDMA

First synthesized in 1912 and patented by Merck in 1914, 3,4-Methylenedioxymethamphetamine, better known as MDMA or ecstasy, was a relatively obscure drug for much of its history.[44] In the 1950s, the CIA dabbled in researching the drug for possible chemical warfare purposes.[45] It wasn't until the 1970s, however, when a group of psychiatrists began to use the drug to facilitate psychotherapy, that recreational use began to grow. Eventually, the drug showed up on the streets. Frightened of a new, unknown street drug with the sexualized name of "ecstasy" that could be used by children, the DEA issued a notice of proposed rulemaking in July 1984, announcing the intent to classify MDMA as a Schedule I controlled substance.[46]

In response to the DEA's proposal, a group of physicians, researchers, and therapists hired DC attorney Richard Cotton to draft a letter to the DEA administrator requesting a hearing on whether and how MDMA should be scheduled.[47] This letter seems to have surprised the DEA, which, according to one DEA pharmacologist, "had no idea that psychiatrists were using it."[48] In fact, MDMA, initially called "Adam" within the therapeutic community (possibly either as a reference to Adam and Eve or as a pseudo-anagram of the letters MDMA), had flown under the radar of the DEA for years. Given that use was usually therapeutic, and that MDMA rarely causes an adverse reaction requiring either medical treatment or law enforcement assistance, it seemed that MDMA could have flown under the radar for years more had it not hit the streets, been renamed ecstasy, and landed in the hands of teenagers. One distributor reveled in the DEA's ignorance:

> One of the wonderful things is MDMA has been known as Adam and used therapeutically in thousands, tens of thousands of sessions for 10 years, since the early, early 70s, when the DEA moved to make it illegal, they had never even heard the name Adam. It wasn't listed at all. It was people who had learned of it from a therapeutic community, some of [whom] had gone on to mass market it under the name of Ecstasy.[49]

An initial hearing occurred on February 1, 1985, in front of DEA administrative law judge Francis L. Young, the same judge who would later rule that marijuana should be rescheduled, only to be overruled by the DEA administrator. Young ordered that three hearings on MDMA be held, one in Washington, DC, one in Los Angeles, and one in Kansas City.

The three hearings produced ten volumes of testimony. Since the DEA was pushing to quickly ban a drug on which there had been little research done, opponents of the ban trumpeted MDMA's benefits as well as the problems that would result from hastily classifying a possibly beneficial drug as "having no recognized medical uses" under the CSA. The DEA, however, was on the warpath, with the assistant administrator, Gene Haslip, telling the *San Francisco Examiner* during the hearings, "We are going to ban Ecstasy within the next several months," because "it's extremely dangerous."[50]

Ron Siegel, a star witness for the DEA, continually mischaracterized MDMA as a "hallucinogen," and told his favorite drug mania stories, including one about a psychologist who decided to direct traffic on a busy street after taking MDMA.[51] The DEA also cited a University of Chicago study showing that MDA, not MDMA, caused brain damage in rats. Although MDA and MDMA are very similar, they are chemically distinct in important ways. Opposing witnesses testified that the two drugs act on the brain in different ways, and that the MDA-damaged rats were administered the drug in highly unusual circumstances. The rats were given extremely high doses of the drug (three to five times a comparable human dose) intravenously rather than orally every 12 hours for two days.[52]

Opponents of the ban presented evidence that MDMA is not addictive, that it rarely results in harmful side effects, and that it had helped thousands of people suffering from various psychological ailments. As such, they argued, the drug should be on Schedule III, not Schedule I. Putting it on Schedule III would allow the DEA to regulate and prohibit recreational use, but it would not keep physicians from prescribing it or hinder scientists from studying it. Schedule III would mean that a possibly truly beneficial drug that had been administered approximately 500,000 times before the DEA had even heard of it could be further studied for methods of safe and beneficial use. Schedule I would essentially be a death sentence for any further scientific research. "I would regard the scheduling of this drug as a scientific calamity," wrote psychotherapist Nathaniel Branden in a letter to the agency. He implored the agency to "leave the door open to further research, exploration, and study in this area."[53]

The DEA, fully ensconced in the drug-warrior mentality of the 1980s, didn't seem to care. In fact, the agency didn't care that it couldn't offer much evidence that MDMA was harmful or that it was being abused. Instead, the agency argued that *actual* harm need not be shown, just the *potential* for harm.

Even though board-certified psychiatrists explained that MDMA had an "accepted medical use" in their practices, the DEA insisted that the FDA should decide what belongs in that category. They argued that the agency "need only ask the FDA whether the drug or substance in question has received FDA approval under the FDCA [Food, Drug, and Cosmetic Act of 1938] in order to ascertain the existence, *vel non*, of 'accepted medical use.'"[54]

In a carefully reasoned 90-page opinion, Judge Young concluded that MDMA should be placed on Schedule III, not Schedule I. Young disposed of the DEA's argument that "accepted medical use" was synonymous with FDA approval. "Congress could easily have linked the phrase 'accepted medical use in treatment' in the CSA to some provision of the FDCA, and FDA's authority thereunder, had it desired to do so. It did not do so." Rather than being determined by the FDA, accepted medical use is determined "by what is actually going on within the health-care community."[55]

Young reviewed "testimony in this record from reputable physicians, i.e., responsible medical authorities who constitute a respectable minority, that the use of MDMA is acceptable in the treatment of certain kinds of patients."[56] Furthermore, the drug, although it can be abused, does not have a high potential for abuse. Finally, reasoned Young, there are accepted safety standards for use under medical supervision, therefore, MDMA should be a Schedule III drug.

Just as he would do with marijuana three years later, DEA administrator John Lawn overruled Judge Young's determination. The action was immediately challenged in the United States Circuit Court of Appeals for the First Circuit, which overturned the DEA's decision on the grounds that the agency's interpretation of "accepted medical use," namely that FDA approval was the sole consideration, was not based in a proper reading of the CSA. "The opportunity for a meaningful hearing would be lost," the court wrote, if the question "turned solely on the existence of FDA approval for interstate marketing." Because the CSA requires an opportunity for a hearing, such a hearing would be "reduced to an empty formality and, for participants like Dr. Grinspoon [who brought the challenge], would amount to an exercise in futility."[57]

Again, in a telling portent of what would transpire with marijuana, the DEA "reconsidered" the question and again decided that MDMA should be a Schedule I drug. In a rule published in February 1988, the DEA gave lip service to the First Circuit's ruling by stating that the "lack of FDA approval" was "not conclusive" in determining whether a drug has accepted medical uses and accepted safety standards.[58] Then, rather than relying on the FDA's determination, the agency merely applied the FDA's standards itself.

The DEA's order perfectly encapsulates the catch-22 situation of trying to prove to the government that a drug meets scientific testing hurdles. This is especially true for drugs such as MDMA, which, because it was discovered in 1912 and is in the public domain, has long since lost any ability to be patented. Pharmaceutical companies, therefore, have little interest in pursuing clinical trials. The DEA nevertheless relied on the lack of clinical trials in determining that safety standards had not been established for MDMA, writing that "very little of this information has been generated for MDMA." The psychiatrists who claimed safe use were using "anecdotal observations" that cannot "substitute for controlled studies in animals and humans." The DEA then obliquely references the rat study, claiming that "there have been studies in animals to show that MDMA produces long term serotonergic nerve terminal degeneration." Because "further testing is necessary prior to human use," MDMA would be placed on Schedule I.[59]

The irony, of course, is that placing a drug on Schedule I is perhaps the best way to shut down scientific research into the drug. The immediate effect of placing MDMA on Schedule I was the "curtailment of scientific research and experimentation with a drug that held therapeutic potential." The FDA and DEA claimed to limit testing and experimentation based on concerns about the health of volunteers, meanwhile recreational use and therapeutic treatment has seen millions of MDMA doses to be taken without the literature showing "even one case of an individual suffering neurological symptoms linked to MDMA-related brain damage."[60]

Five applications were submitted to the FDA to research MDMA between 1986 and 1989, and all were denied. According to MAPS, "the FDA based its rationale for rejecting all protocols and single case studies on the hypothetical risk of functional consequences of potential neurotoxicity from MDMA." Things began to improve when, in 1992, the FDA approved a phase 1 study on MDMA use in alleviating pain, anxiety, and depression in cancer patients. That study showed that MDMA posed no unusual risks and could be safely administered. Soon after, though, the FDA put "MDMA psychotherapy research on a slow track to nowhere."[61]

Funding is another issue. Government funding for research into psychedelics is almost nonexistent. As Tom Insel, the director of the National Institute of Mental Health (NIMH) told journalist Michael Pollan, "The NIMH is not opposed to work with psychedelics, but I doubt we would make a major investment."[62] Funding for research has most often come from private donors like the Beckley Foundation in the UK, the Heffter Research Institute, and MAPS.

In the limited research that has been done, however, MDMA has continually surprised researchers with its efficacy in treating certain psychological problems, particularly PTSD. In one small study of those with treatment-resistant PTSD, 80 percent of MDMA-treated patients reported benefits from the treatment; only 20 percent of the placebo group did so.[63] Even one year later, the majority of those treated with MDMA reported continued beneficial effects, whereas none in the placebo group did. In another small study of those suffering chronic PTSD, 10 out of 12 subjects were determined to be cured after two MDMA-assisted psychotherapy sessions.[64]

David Nutt, the aforementioned British pharmacologist, was dismissed from the British government's Advisory Council for the Misuse of Drugs when he publicly suggested that horseback riding is more dangerous than ecstasy (it is). In his book *Drugs without the Hot Air*, Nutt put the psychotherapeutic allure of MDMA in stark terms:

> If we wanted to invent a drug especially designed to help enhance trauma-focused therapies, it would have the following qualities:
>
> 1. Be short-acting enough for a single session of therapy.
> 2. Have no significant dependency issues.
> 3. Be non-toxic at therapeutic doses.
> 4. Reduce feelings of depression that accompany PTSD.
> 5. Increase feelings of closeness between the patient and therapist.
> 6. Raise arousal to enhance motivation for therapy.
> 7. Paradoxically, increase relaxation and reduce hypervigilance.
> 8. Stimulate new ways of thinking to explore entrenched problems.

Ecstasy has all these qualities when used in a clinical setting, and it is extremely effective.[65]

Since then, research into psychedelics, particularly MDMA, seems to have turned a corner. In November of 2016, the FDA approved a large-scale phase 3 clinical trial of MDMA-assisted psychotherapy,[66] and, in August 2017, two more phase 3 clinical trials were approved.[67] What's more, the FDA designated MDMA-assisted psychotherapy as a "breakthrough therapy" and thus will work to complete the final phase quickly. This is primarily due to the astounding effectiveness of the phase 2 trials, like previous research:

> In MAPS' completed Phase 2 trials with 107 participants, 61 percent no longer qualified for PTSD after three sessions of MDMA-assisted psychotherapy two months following treatment. At the 12-month follow-up, 68 percent no longer had PTSD. All Phase 2 participants had chronic, treatment-resistant PTSD, and had suffered from PTSD for an average of 17.8 years.[68]

Providing the funding for the studies is MAPS, which is "currently the only organization in the world funding clinical trials of MDMA-assisted psychotherapy."[69] Through the effort of MAPS and other organizations, we are on the cusp of seeing the approval of a bona-fide breakthrough drug for PTSD treatment. But if 40 years ago the DEA had listened to the psychologists who touted the benefits of MDMA, untold numbers of PTSD sufferers could have been helped.

The growing acceptance of research into MDMA has also helped resuscitate research into other psychedelics that were hastily banned and given a bad reputation before any meaningful scientific research could be done, particularly LSD and psilocybin. Psilocybin has been used to combat smoking addiction in small, FDA-approved studies with shocking results. In a study with 15 participants, "twelve subjects, all of whom had tried to quit multiple times, using various methods, were verified as abstinent six months after treatment, a success rate of 80 percent." Before it was stamped with the label "hippie drug" and associated with anecdotal stories of "bad trips," LSD was used to successfully treat alcoholics.[70] But getting into the story behind psilocybin and LSD is beyond the scope of this essay. Suffice it to say that the story is largely the same as for MDMA: a fearful government, goaded on by hyperbolic media reports and frightened parents, banned something before we knew much about it. As a result, science was retarded, and those who could have benefited were forced to languish with their possibly curable disorders.

Conclusion

Drugs, especially "new" drugs, scare people. This is understandable. But that fear should not be allowed to divert public policy from its proper moorings in good science. Prohibition is sometimes thought of as the ultimate form of regulation. But this notion is mistaken. Prohibition is the absence of regulation, either for street use or for scientific studies. Prohibition means anarchy.

Placing a drug on Schedule I virtually guarantees that our knowledge about the drug's effects and possible beneficial uses will be hampered. In the process, the true victims will be those who could have benefited from using the drug, but instead must live in a world created by the fears and trepidations of past generations. We seem to be turning a corner on marijuana and MDMA, and it can be hoped that we've learned some lessons about the costs of knee-jerk reactions to "drug scares."

MEDICAL INNOVATION AND THE

GOVERNMENT-ACADEMIC-BIOMEDICAL COMPLEX

The incentive structure for advancement in the biomedical sciences differs very little from that in other fields. Success or failure is judged through the prism of academic publications generated by outside financial support—in this case, both public and private. Here we discuss the concept of innovation in medicine and its relationship to the reward structure. We will find that the two are almost completely independent of each other. In other words, federal funding for research is not very related to progress in patient care, in the same way that public funding and private funding produce roughly the same amount of innovation in basic and applied research.

"Medical science," a widely used term, is misleading. Without question, the study of the natural sciences has contributed to the fact we lead longer and healthier lives than previous generations. But it is *technology*—the availability of practically useful drugs, diagnostics, and medical devices—that deserves most of the credit for these improvements. The failure to recognize that distinction is the basis of a common misconception that free-ranging research by academic scientists—funded predominantly over the past half-century by the federal government—is the main basis of medical advances. This chapter reviews the historical reasons why this idea is false.

The First Era of Medical Innovation

Substantive medical progress emerged in an innovation era that began in the middle of the 19th century. Fueled by the capital of the Industrial Revolution, medical research proliferated. Discoveries by Louis Pasteur, Robert Koch, Ignaz Semmelweis, John Snow, Joseph Goldberger, Walter Reed, and many others gradually ushered in a range of technologies, including vaccination against infectious diseases, antisepsis, anesthesia, sanitation, treatment of vitamin deficiencies, and elimination of insect disease vectors, all of which contributed to the prolongation of life.

In the early years of the first innovation era, these discoveries often emanated from state-funded European universities and research institutes. Subsequently, a few private institutions in North America such as Johns Hopkins Medical School and the Rockefeller Institute engaged in medical research supported by private philanthropy. Their work was later supplemented by associations dedicated to specific maladies such as heart disease, rheumatism, and cancer. Medical innovation also arose from drug discovery generated by partnerships between academic physicians and the German dyestuffs industry. These led to aspirin to treat pain and fever, and to antimicrobial therapies such as arsenicals (for syphilis and parasitic disease), sulfonamides, penicillin, and streptomycin.[1]

In America, until early in the 20th century, proprietary apprenticeships competed with and overshadowed the European tradition of university-based education of physicians. The 1910 Flexner Report, however, enabled organized medicine, eager to restrict professional access and reduce competition, to collude with universities to have states only license physicians who had graduated from accredited academic medical schools.[2] An important criterion for academic advancement in these institutions was the performance and publication of research. This resulted in election of medical researchers to elite professional societies such as the Association of American Physicians, the American Society for Clinical Investigation, the Interurban Clinical Club, and the Clinical and Climatological Society that convened gatherings to share research results. Sinclair Lewis's Pulitzer Prize–winning novel *Arrowsmith*, published in 1925, along with other contemporary accounts, describes how doing research in medical schools conferred prestige on its practitioners and the tension between research and nonresearch physicians who felt demeaned by them.[3]

Although the number of American medical researchers was relatively small during the first innovation era, they made important advances, such as treatments for pernicious anemia, hormone replacement therapies, and surgical treatments for heart valve disorders or to remove blood clots from occluded vessels.[4] During World War II, academic medical researchers' contributions to the war effort resulted in the development of anti-malarials, mass production of antibiotics, and blood transfusion therapy.[5]

The Second Era of Medical Innovation

The role of technological superiority as a decisive factor in the Allied victory was a transformative element that ushered in the current era of medical innovation. The principal architect of that transformation was Vannevar Bush (1890–1974). An engineer, entrepreneur, and academic administrator, Bush came to head the entire research-and-development organization of the American war effort, culminating in the Manhattan Project and the atomic age.[6]

At the behest of Franklin Roosevelt, Bush wrote a treatise—*Science: The Endless Frontier*—making the case for lavish government funding for research.[7] A central element of Bush's plan was the idea that basic research, or research for research's sake—"the free play of free intellects"—was the best path toward progress. The emerging Cold War, with its mandate for achieving nuclear superiority, was a prime mover in ensuring a political embrace of Bush's vision. But on the medical front, influential university officials and researchers invoked the recent success of a vaccine to combat frightening polio epidemics and other achievements to argue for applying Bush's ideas to medical innovation. Congress enthusiastically complied with a massive buildup of the National Institutes of Health (NIH).

The NIH's predecessors originated in 1798 with the Marine Hospital Service to care for merchant seamen (the NIH logo still features an anchor!). It evolved into a public health organization, the Hygienic Laboratory, to address infectious diseases and in 1930 acquired the title of NIH (Institute in the singular), which expanded to include multiple institutes during World War II.[8]

As an ultimate result of Bush's proposal, through postwar congressional initiatives, the NIH began receiving much larger appropriations.[9] It constructed a laboratory research facility and a hospital dedicated entirely to research on a spacious tract of land in the Washington, DC, suburb of Bethesda, Maryland. Most of the NIH's funds, however, began to flow "extramurally" to universities and independent research institutes. Thus began the second era of medical innovation, giving birth to what is best called the government-academic-biomedical complex (GABC), which remains the dominant bankroller of medical research in American academic medicine.

The Growth of Academic Medical Research

The premium on research as the route to advancement in academic medicine conspired with the availability of money to recruit more and more faculty members to that endeavor. Medical schools not previously counted among the elite achieved sudden prominence. The University of Texas-Southwestern in Dallas evolved in the late 1940s from a collection of former military barracks into a major academic powerhouse, culminating in the 1985 Nobel Prize in Physiology or Medicine awarded to Michael Brown and Joseph Goldstein for their work on cholesterol metabolism.[10]

The Korean and Vietnam Wars' drafting of physicians in postgraduate training into the military additionally contributed to the medical research workforce.[11] Those with academic ambitions could fulfill their military obligation in the Public Health Service and obtain positions at the NIH (or at the newly founded Centers for Disease

Control) facilities to receive research training. Competition for these slots—facetiously named "the yellow berets"—was stiff and resulted in the rapid population of the NIH "intramural" Bethesda facility with young trainees, many of whom became career medical investigators.[12] The emphasis on research led medical specialties' accreditation boards to stipulate research training as a requirement for specialist certification. Physicians training to be specialists had to spend time in research programs whether or not they wanted to. The NIH provided stipends for such training both to academic departments and to individual trainees, who were cheap labor for university research programs. NIH funds were also made available to universities to encourage undergraduates to obtain research experience and to subsidize graduate training programs.

Money went begging in the startup phase of GABC-fueled extramural granting. NIH officials solicited universities for research grant applications. But this vacuum filled rapidly. Medical school departments, such as surgery, that had traditionally done little research, soon initiated laboratory-based research programs. Since surgeons spend so much time in the operating theater, they hired basic scientists with PhD degrees to work in their laboratories. The emerging research programs needed expensive equipment, glassware, chemicals, and experimental animal facilities. Most heads of research laboratories spent their time reading research papers and designing experiments, leaving the execution of those experiments to researchers in training (graduate students, interns, residents, and technical assistants). Grant applications and research papers—all typewritten in those days— required secretarial support. The NIH underwrote it all. In the late 1940s, the American Cancer Society was the largest funder of medical research. By the mid-1950s, its contributions lagged far behind the government's.

An especially attractive feature of NIH funding for universities and research institutes was that it provided much more generous indirect (overhead) support than private philanthropists, who preferred to see their donations underwrite research and researchers rather than institutional infrastructure such as building construction and maintenance and administrative costs.

This bounteous federal research funding party lasted only about a decade and a half, up to the late 1970s. Congressional NIH appropriations did not decrease thereafter but rather, as discretionary budget items, could not rise in proportion to the demands by academic centers for more and more externally funded research. Despite general congressional enthusiasm for the NIH mission, the Johnson administration's Great Society programs and Vietnam War spending precluded further expansion of the NIH. Congress and the NIH's management began to impose restrictions on the use of NIH funds, such as disallowing payments for secretarial services and requiring training grant recipients to spend post-training time in bona-fide academic research programs or else pay back the funds they had received.

Financial constraints increasingly became a serious problem within the GABC. But even beginning with the first signs of fiscal distress, the GABC's beneficiaries—universities and their guilds, research institutes, professional societies, patient advocacy groups, and voluntary health agencies—lobbied Congress strenuously for more NIH funding. Unsurprisingly, their efforts at persuasion centered on the premise that such subsidy is essential for medical innovation.

This plausible hypothesis is wrong. Whereas others have provided cogent critiques of the role of government in science and of Vannevar Bush's enthusiastic promotion of basic research, this analysis drills down more specifically on the problem of *medical* innovation, which has—I believe—unique features that differentiate it from other research-based endeavors.[13]

Medical Innovation versus Basic Research in the GABC

Assessing contributions to "medical innovation" demands a precise definition of that term. Unfortunately, innovation gets muddled with novelty. Clearly, novelty has its benefits—most palpable in the arenas of fashion and entertainment. What matters in healthcare, however, is what improves longevity and life quality by promoting health and combating disease. The definition of medical innovation should be restricted to the technologies that accomplish those ends. Innovation requires the intermediary step of having a research result suggest a practical possibility, an outcome I define as "invention." When an invention comes to practical use, innovation occurs.

Attesting eloquently to the divide between GABC-based research and innovation is an examination of 25,000-plus publications in prestigious biomedical research journals. Only 100 even *mentioned* invention, a practical implication of the reported findings.[14] Also telling was the passage in 1980 of the Bayh-Dole Act to encourage universities to license technologies to companies for practical development, prompted by the observation that discoveries were either not being patented or patents were not resulting in licensing.[15]

Almost three decades after Bayh-Dole was enacted, university income from all licensed inventions—not just medical innovations—remains a minor sum in comparison with the totality of healthcare revenues—or for that matter university revenues, being less than three billion dollars a year.[16] In fact most universities lose money on innovation efforts because their revenues do not even cover the administrative costs of licensing.[17] Clearly government funded basic research is not generating a significant number of useful medical innovations.

Unquestionably, medical innovation proceeded apace during the early GABC era, but not at any increasing rate, though funding for basic research did grow substantially. A benchmark for innovation is the regulatory approval of drugs and medical devices. The average number of such approvals has been constant for the past half-century despite the introduction and continuous operation of the GABC over that interval.[18] Moreover, independent analyses have concluded that over 80 percent of the drug approvals arose solely from research and development in private industry.[19]

The Revealing Case of Heart Disease

In addition, an examination of some particular innovations during the GABC era raises questions regarding the GABC's specific contributions. Over the first half of the 20th century, deaths from heart attacks and strokes had risen steadily, reaching their highest peak at the GABC's onset in the 1950s.[20] Today, six decades later, cardiovascular mortality has fallen by over 60 percent.[21]

Many factors account for this improvement. Risk factors for vascular disease, such as smoking, high blood pressure, abnormal blood cholesterol levels, and genetic predisposition were unknown in the middle of the 20th century; research identified them and led to efforts to mitigate them. But does the GABC deserve most of the credit for this accomplishment?

A principal source of cardiovascular disease risk identification was the NIH-sponsored Framingham Study that correlated the existence of heart disease with the presence of such factors in the population of the Massachusetts city for which the epidemiologic project was named.[22] Although the Framingham Study began before the extensive expansion of the GABC, it clearly deserves recognition for its contributions.

How this knowledge translated into innovation is complex. When the Framingham parameters associated with heart disease emerged, whether they were causal or merely correlative was unclear. In fact, other NIH-funded epidemiologic studies that purported to link dietary fat intake to heart disease were ultimately deemed flawed and uninformative.[23]

Medications—produced by companies—to treat high blood pressure existed before recognition of its association with heart disease. Their principal use was to combat "malignant hypertension," extremely high pressures that acutely damage the heart, brain, and kidneys, because these drugs were too toxic to apply to milder cases of hypertension. Once companies developed more tolerable blood-pressure-reducing agents, large population studies demonstrated that lowering blood pressure diminished the incidence of heart attacks.[24]

Similarly, whether elevated blood cholesterol levels truly caused heart attacks and strokes was also a topic of debate. Only the development of cholesterol-lowering "statin" drugs—an effort solely driven by industry research—established that high blood cholesterol is a bona-fide heart disease risk factor and that reducing it has preventive value.[25]

Another key contribution to the reduction of heart disease mortality has been the introduction of imaging techniques to diagnose narrowing and closure of arteries that feed blood to the heart and of devices and drugs to restore blood flow. With the exception of the clinical trials that have established the effectiveness and safety of these interventions, the preponderance of effort underlying these innovations took place in private industry with the important input of entrepreneurial physicians. A similar argument applies to the development of joint replacements, which have markedly mitigated the pain and immobilization inflicted by arthritic conditions.

The Case of the Genetic Revolution

Space limitations preclude examining the universe of innovation that has appeared during the GABC era. But worth mentioning is the highly touted genetic revolution. This effort spawned the biotechnology industry that accomplished the discovery and application of potent "biological" therapies. It also generated the "genome project" that elucidated the sequences of the human genes and those of other species. One oft-cited example of seemingly impractical "basic" research that strongly abetted this genetic technology was the discovery of "restriction enzymes," compounds produced by bacteria that cut up DNA in specific ways that enable researchers to define and manipulate genes.

Though unequivocally useful, work on restriction enzymes and other discoveries leading to elucidating the structure of DNA was subsidized by sources such as the American Cancer Society rather than the government long before the buildup of the GABC.[26] And by credible accounts, although the GABC initiated the genome sequencing project, without its rescue by what started as competition from the private sector, its completion would have been markedly delayed.[27]

These examples—especially the decoding of the human genome—contradict the passionate advocacy for the GABC's essential role in medical innovation by its academic beneficiaries. The naiveté of the facile assumption that basic research is the progenitor of innovation is also discussed in Chapter 1 by Kealey and Michaels.

As concerns began to arise in academic institutions regarding the sustainability of government funding for basic research, two physician-researchers, Julius Comroe and James Dripps, attempted to counter an inconvenient finding that emerged in the 1960s. A study dubbed Project Hindsight had examined weapons technology innovation and posited that the vast predominance of progress in weaponry was the result of work directly related to the subject at hand, not basic research.[28] In an influential paper published in the journal *Science*, Comroe and Dripps came to the opposite conclusion: that advances in the treatment of cardiovascular disease management—their area of clinical specialization—originated from research findings with no direct connections to the ultimate technological achievements.[29]

But rather than meticulously documenting the steps involved in the evolution of specific military technologies as was done by Project Hindsight, Comroe and Dripps relied on the indirect approach of having "experts" opine about the provenance of medical innovations, an exercise fraught with subjectivity and potential inaccuracy.[30]

Innovation is inherently nonlinear and routinely depends on adaptation and serendipity to exploit findings applicable to practical problem solving. Although some such results arise from undirected basic research, this fact does not in and of itself justify a massive investment in research for research's sake. The Comroe-Dripps compilation also failed to quantify the large denominator of research results that contributed nothing to innovation, an important element of any cost-benefit analysis.

A more recent attempt to give the credit for drug discovery to university research also recruited "experts" to opine on what they considered to be the "most transformative" drugs developed between 1984 and 2009.[31] The authors concluded that the most impactful drugs arose from academic research.

It is logical that communities that have benefited so much from the GABC will fiercely proclaim its virtues, even resorting to substantial rhetorical and logical leaps. Like the Comroe study, the more recent experts' drug selections also involve subjective opinions, and the focus on "transformative" drugs reflects the academic obsession with what is novel and impressive. In the real world, any given patient wants a drug that *works*. A medication that relieves a headache effectively is "transformative" for that individual. Also, most of the drug approvals cited in that study took place early in the study interval, too soon for the GABC to have had a major influence.

Finally, and most important, the study's authors' casual anecdotal descriptions of drug development were skewed—often inaccurately or inappropriately—to overemphasize the importance of academic contributions. For example, the narrative implies that a highly effective anti-leukemia drug, inatimib, was discovered by an academic physician grounded in basic science principles.

The truth is that a company discovered the drug looking to treat diseases other than leukemia, and while the academic researcher's contribution was material, the drug never would have helped patients in the absence of industry efforts.[32] The article reporting the study neglected to cite most of the previous research that more meticulously analyzed the development histories of important drugs. Most telling, a separate and far greater in-depth analysis of the provenance of the same "transformational" drugs covered in the study ascribed the dominant influence to the private sector.

The Flawed GABC Incentives

If the march of medical innovation has not benefited from the GABC as much as advertised, what could be the explanation? The principal answer is misaligned incentives. If history teaches anything, it is that incentives and rewards work far better than good intentions, especially those dictated by authorities. The GABC's incentive misalignment relates to the definitions discussed above: the GABC rewards "research" more than it does "innovation."

Producers of goods and services dependent on revenues must innovate in order to survive. Because sales monopolies enjoyed by manufacturers of brand pharmaceuticals and medical devices are temporary, a thriving enterprise producing follow-on generic products mandates that the innovative brand industry maintain a pipeline that

addresses previously unmet medical needs or provides better management than what is currently on the market.

Researchers outside of industry have been happy to advertise that they
have high-minded intentions—improving mankind's lot by increasing
knowledge and providing practical innovation. This propaganda impulse is understandable in view of the fact that such researchers depend heavily on public philanthropy for their livelihoods. To unlock
that purse, researchers have created the impression that they promote
material progress by operating in a universe of higher—"objective"—
reality than purveyors of mere business. The opacity of science to
most nonscientists plays into that message.

But historians of science have documented that researchers work
predominantly for recognition of their accomplishments by influential peers, and that they fight tooth and nail for that attention.[33]
Nothing attests more powerfully to this conclusion than the profound importance the establishment of scientific journals in the
17th century had on the advance of science.[34] By providing an institutional repository for scientists to claim credit for discoveries, journals overcame scientists' reluctance to publicize their work for fear
of intellectual pilferage.[35] In addition, scholars have challenged the
conceit that science reveals absolute, "objective" "truths." Rather, it is
a very social activity that depends on a very mutable interpersonal
consensus influenced by personal preferences, career ambitions, and
political power.[36]

A prime example of the different incentives for "medical innovation" and academic "medical research" is that the former receives its
rewards for achieving regulatory approval of new products for clinical use. The latter seeks recognition through publication of research
papers, awarding of research grants, academic promotion, admission
into scientific elite organizations, and receipt of prizes. If these academic rewards accrued to researchers because they promoted innovation, the GABC's incentives could be appropriate. Instead, historians
and sociologists of science have concluded that academics value and reward elegant solutions for complex scientific puzzles more than they
do innovation that saves or enhances life.[37]

Interesting versus Useful

One articulation of what constitutes value in science lists "accuracy, intrinsic interest and general relevance."[38] Ironically, whereas the criterion of "accuracy"—also definable as reproducibility—would seem to go without saying as an essential criterion of value, the bar set for academic research is that observations need only achieve minimal statistical validity without independent corroboration to justify publication. It is the norm that companies attempting to exploit academic findings as possible product candidates are often unable to replicate the academic results.

The criteria of "intrinsic interest and general relevance" raise the question: of interest and relevance to whom? Experience reviewing research papers submitted to journals for publication and grant applications reveals that new ("novel") and unexpected findings elicit maximal interest. General relevance, such as laws that predict the behavior of moving objects, commands the highest adulation. But notably, practical utility is not on the list. In the case of biomedical science at the outset of the GABC era, Vannevar Bush's lionization of basic research played seamlessly into this value system.

The Case of Reductionism versus Innovation

Even as medical innovation accelerated in the first, pre-GABC, innovation era, medical practitioners had little or no idea how most of their beneficial tools worked. Trial and error identified malfunction of the pancreas as a cause of diabetes and subsequently the existence of insulin, the hormone it produced, and the deficiency of which resulted in the disease. Even then, in keeping with the fact that only industry had the wherewithal to exploit such information for innovation, the University of Toronto researchers who discovered insulin tried and failed to do more than demonstrate its clinical efficacy to treat diabetes in a few patients. Insulin therapy did not have a major clinical impact until the Eli Lilly Company took over the problem and mass-produced the hormone.[39]

But as the GABC expanded and new technologies for analyzing the structure and function of body components and subcomponents at ever more fundamental (molecular) levels came on line, the plausible but empirically largely unsupported idea that more detailed knowledge concerning how derangements of these processes cause disease and how addressing them produces cures held sway. The reductionist science that had so profoundly advanced physics and chemistry had great appeal for the biologists previously mired in empiricism—and more and more physicists and chemists seduced by GABC resources moved into basic biological science.

The reductionist paradigm, warmly embraced by the GABC, creates endless research questions, few of which, as recognized by others, lead to medical innovation and progress.[40] One aspect of the primacy of reductionism is an obsession with "molecular mechanisms." The fact, for example, that aspirin reduces fever became less compelling than the "molecular basis" of *how* it happens.

Bush Was Right: Research Becomes Endless

Perversely, the expansion of the menu of "intrinsically interesting" phenomena unearthed by basic research over time presents far more opportunities for researchers to follow up with adequate depth or breadth. Instead of the finite endpoint of approval of a practical innovation by regulators persuaded that something works with acceptable risks, the basic research system accommodates open-ended wheel spinning. Editorial commentary concerning research published in journals inevitably concludes that "more studies are necessary," as if a bottomless supply of researchers exists to advance the knowledge frontier closer to a never-receding illusory goal. This attitude brings to mind the scenario described by Herman Hesse in his 1943 novel *The Glass Bead Game*, in which abstract cerebral virtuosity divorced from any real-world activity has become the ultimate end of intellectual activity.[41]

For all its pretensions to research that abets innovation, the GABC ignores and dismisses a vast repertoire of essential but academically unsung skill sets required to achieve innovation. One of hundreds of these tasks is "formulation." A drug candidate may work spectacularly in a test tube to, say, kill cancer cells, but if it cannot be prepared in a form amenable to absorption by the body following oral or parenteral administration that persists long enough to get to the body sites where its action is desired and do what it is supposed to do, its test tube performance has no practical utility.

Economic Impacts

Separate from whatever damage the GABC's emphasis on basic research has inflicted on medical innovation, its economic impact has arguably compromised it. Although compelling arguments—usually citing national defense—justified the postwar infusion of government support's transformation of universities from bucolic bastions of predominantly ivory tower contemplation into more diverse institutions contributing practical inventions as well as abstract knowledge and educated citizens to society,[42] the financial dependence on government support that accompanied this transformation inflicted deleterious consequences customarily associated with addiction.

One way to track whether the GABC kept pace with innovation opportunity is to compare its investment with the accrual of individuals living beyond 65 years of age, a figure easily obtained from census data.[43] The comparison reveals that both figures rose proportionately, and exponentially, until the late 1970s when, as NIH appropriations remained essentially flat, the longevity curve diverged steadily upward away from it.[44]

This discrepancy raised two issues, one stridently engaged by the academic community, and the other ignored by it. The one that received attention was the fact that the NIH appropriations were rising more slowly than in the past, predictably lowering the success rate of grant applications. Academic officials raised alarms about the effect this austerity was having on the recruitment of physicians into medical research, thus encouraging a diaspora of physician-researchers from research activities into clinical or administrative efforts.[45] The ignored issue was to ask why population survival rose unabated despite the constrained GABC investment. The obvious answer was that private investment continued to increase and, in fact, has grown precisely in parallel to the data on aging.[46]

In response to slowing growth of NIH funding appropriations during the 1980s and early 1990s, academic distress escalated. Biomedical research societies and voluntary health organizations exhorted their adherents to lobby Congress for greater generosity. The arcane nature of research, the ineptness of the biomedical community in explaining it, and the competing political agendas of welfare, other entitlements, and defense made for a difficult challenge. Except for a brief five-year period in the mid-1990s during which Congress doubled the NIH budget appropriation, the gap between public and private investment in biomedical research has continued to widen.

The financial hardship in universities dependent on external research subsidies and other developments altered the demographics of the academic medical workforce. The earlier lamented decline in the number of physicians involved in research accelerated markedly.[47] For the most part, the only physicians seeking research careers had also obtained PhD degrees during an overly long training sojourn—often doing so to take advantage of tuition subsidies rather than a deep commitment to research, as evidenced by a high post-training dropout rate.[48] Hence, individuals with PhDs but no medical training came to dominate the research effort.

The researchers who had obtained faculty positions and research support defining them as "principal investigators" depended heavily on NIH salary support, because many academic institutions had recruited such individuals without setting aside funds to pay their salaries. In an effort to counter this trend and to increase the efficiency of research grants by having them cover research, equipment, supplies, and technical assistants rather than subsidizing principal investigators, the NIH capped salary budgets. One result of these developments has been that academic tenure, at one time a sinecure, has become relatively meaningless in research-intensive universities. Principal investigators unable to obtain NIH grants leave research for medical administration, clinical practice, or industry.

The average age of researchers able to obtain grants has increased markedly because only applicants with established track records and in possession of an entourage of trainees and younger colleagues can generate enough new data that may rise to the level of intrinsic interest for sufficiently favorable review opinions to surmount the high funding priority bar.[49]

Paradoxically, the GABC's straitened financial circumstances have undermined the cliché that undirected basic research addresses wide-ranging long-term scientific goals in contrast to commercial research and development allegedly committed to immediate problem solving and fast profits. Faced with ever-shortened NIH funding intervals (now three to four years), GABC supplicants must operate with narrowly focused efforts tightly linked to previous accomplishments in hopes of surmounting the aforementioned bar.

The prospects for the successful principal investigators' entourage rising to the status of their superiors are vanishingly slim; they may orbit from one postdoctoral fellowship to another until they give up and leave academe for research jobs in industry—which are finite—or teaching positions or some other endeavor.

One might imagine that the leaders of academic institutions would be in the forefront of advocacy for more government research funding. Although these managers articulate such sentiments, their major concerns, reflected by the principal lobbying emphasis of the Association of American Medical Colleges that represents them, is the continuation and, if possible, expansion of government subsidies for clinical training of interns, residents, and fellows who, as a result of this welfare, represent cheap labor for providing medical services.

Left to fill the vacuum of special pleading for the GABC is advocacy entrepreneurship engaged by an organization called Research!America. Funded to some extent by universities but principally by the pharmaceutical industry, Research!America relentlessly exhorts biomedical researchers to plead with their legislators for more NIH funding. But what it does best is collect celebrities for gala events, conferring awards on them and on high-paid biomedical bureaucrats. Although absent this group's efforts things might have been worse in theory, NIH funding adjusted for inflation has declined since its founding in the 1980s.

Of all of its counter-innovative elements, the GABC's worst is arrogance. The principal manifestation of this hubris is that GABC has inverted the reality that it is private enterprise: it is the pharmaceutical and medical device industry, not the GABC, that has been accountable for most medical innovation over the past half-century.

The principal instigators of this misconception are entrepreneurial academics who have built careers not only by exploiting public ignorance of the facts by overstating the contributions of NIH-sponsored research but also by engaging in a wholesale demonization of the industry as corrupt. I refer to this activity as "pharmaphobia";[50] a signature characteristic of its acolytes is that none have made substantive contributions to medical innovation. Their critical narrative plays into the hands of sensation-seeking media, populist politicians, and predatory litigators who profit from charging real or imagined corruption.

The idea of bias induced by commercially driven funding has become a separate area of research in the GABC world. Commercial funding is viewed with suspicion, as somehow intrinsically evil, despite the fact that this is where medical innovation comes from. That government funding might be even more pernicious is not ever considered. This discriminatory research supports pharmaphobia, creating a guilt-by-association barrier to making academic research useful. So does the extensive use of so-called conflict of interest disclosure requirements. Here the implication is that any funding from a commercial firm somehow taints the research. This makes it very hard for GABC researchers to work on innovation.

Medical journal managers have enthusiastically joined the pharmaphobia bandwagon because it allows them to sell the false idea that their periodical products embody the most reliable information, which they improperly contrast with pharmaceutical marketing. In fact, as the poor reproducibility of published results attests, the vetting of industry's research and development work by professional regulators is far more rigorous than the relatively casual oversight of research published in journals, and companies can only publicize what the regulators have approved. The overwhelming majority of published journal articles retracted because of fraud are of academic origin and have no industry association.[51]

In addition to discounting the value contributions of industry, the pharmaphobia narrative denies the difficulty and expense of innovation that has increased 80-fold since the onset of the GABC.[52] The origin of this rise is tightened regulatory requirements engrafted on the fundamental unpredictability of biology. On the positive side, biological irregularity is what has enabled us to survive against trillions of rapidly evolving microorganisms that utilize our body components to sicken us. The downside of this adaptation is that more than nine out of ten drugs that seem promising in laboratory experiments or inbred animals fail in clinical trials, and therefore the one success must pay for the nine losers.[53]

This grim economic reality mandates that the industry must achieve sufficient profitability to sustain innovation. The pharmaphobia narrative's misguided denial of this fact underlies calls by poorly informed activists to confer price controls on American drug manufacturers.[54] American consumers, who live in the only country lacking such controls, subsidize the entire world's medical innovation.[55]

The pharmaphobia narrative detracts from medical innovation in many ways. Regulations it has spawned in academic health centers and in state and federal governments, designed to root out speculative corruption, at worst complicate, delay, or outright prevent relationships between academic researchers and industry that historically have promoted innovation. At a minimum, they divert scarce resources from actual innovation to bureaucratic management.

The most extreme example of this dysfunctional regulation is "sunshine legislation" embedded in the 2010 Affordable Care Act, which mandates that companies report to the government payments ("exchanges of value") of cash or in kind exceeding ten dollars to licensed healthcare practitioners for public dissemination. The reporting (at a cost of hundreds of millions of dollars) has predictably had no impact on patient care or patients' trust in their healthcare providers, but its potential for shaming has steadily increased, to 36 percent, the proportion of clinical practices that refuse to give company representatives access to their practitioners, thereby depriving physicians of reliable medical information. What is particularly unfortunate about the pharmaphobia narrative is that it stands in the way of reconfiguring the GABC to make it more compatible with medical innovation.

Reforming the GABC

According to Socrates, "admitting ignorance is the beginning of wisdom." The GABC equivalent of this insight is to jettison the "bench-to-bedside" legend that academics promote, claiming that basic research reliably generates discoveries that seamlessly flow down the development pipeline into inventions that become innovations. For the most part, GABC behavior remains at the bench, piling up often nonreproducible and nontranslatable reductionist findings. Here, briefly, are some reform options:

1. Supplant the legend with the reality of how innovation actually occurs. The truth is that, almost invariably, innovations come to pass in fits and starts, overcoming myriad obstacles and succeeding only if moved forward by committed champions.[56]

2. Exploit the competitive advantage conferred by the fact that many healthcare workers are more interested in *care* than in research. The fact is that most healthcare providers will never be excellent researchers or innovators because the work does not interest them. Recognition of this fact could stop the venerable but increasingly anachronistic tradition of force-feeding healthcare trainees with "scientific" information that has little relevance to everyday practice and could shorten healthcare professional training and reduce the medical school science faculty population and the financial burden it creates.

Such an attitude change could also nudge the emphasis in medical schools toward practical problem solving, especially if promotion and recognition rewarded it rather than the publication of arcane papers in prestigious journals and acquisition of basic research grants as the vehicle for advancement. Since problem solving is the major goal of the medical products industry, such an attitude change could set the stage for industry to shore up the declining financial contributions to research by the GABC.

3. Understand that industry should not and will not simply supplant the GABC model of subsidizing research for research's sake. Basic research has value and will persist in academia, though, as Terence Kealey explains in Chapter 2 in this volume, it does not require massive public funding. In addition, downsizing the basic research enterprise to a level that academic institutions can responsibly and sustainably support would make a far more persuasive case for some public supplementation than arguing that government should be almost wholly responsible for academic biomedical research as an entitlement.

4. Accept that the cultural elements that reward the pursuit of intrinsic interest and general relevance in scientific endeavor seem to be hard-wired and will no doubt persist among basic researchers. However, academic officials can help to level the playing field by promoting and honoring individuals who engage in the efforts required for innovation. Successful prosecution of intellectual property and licensing of technologies to industry should have as much cachet as publication in high-profile journals and awarding of grants.

The NIH currently supports such activities through its Small Business Innovation Research (SBIR) program. A problem with this system is that (predictably) the NIH tends to convene review panels that emphasize basic science expertise rather than the technological competence required to promote innovation proposed by the projects. The result is that these grants receive low ratings and are funded only because the number of applications is far lower than the number of standard research grant requests.

These changes would ameliorate the pharmaphobic attitude and perhaps raise the possibility that private industry will take a serious interest in partnering with academia to promote innovation. This might actually make the GABC an engine of innovation.

In a slim volume published in 1952 entitled *The Cost of Health*, Ffrang-con Roberts, a British physician, documented an exponential rise in health spending in the United Kingdom (since 1900) and correctly ascribed it to the fact that instead of dying prematurely of previously unmanageable diseases, many people were benefiting from what he eloquently characterized as "medicated survival."[57] Although the expense figures Roberts compiled were a mere blip on the ascending cost curve that followed, its upward trajectory has not changed since.

Roberts concluded that the only possible remedies for the Malthusian expansion of medical costs were to ration healthcare or for a nation's economic productivity to grow sufficiently to accommodate the costs. Over the past decade, relatively slow economic growth has rendered this second alternative extremely challenging.

The implication of Roberts's analysis was that medicated survival axiomatically increases healthcare costs. If these costs exceed society's ability to sustain them, is investment in medical innovation ultimately, as some aver, detrimental to society?[58] A difference between Roberts's 1950s perspective and today's is that medicated survival is now far more compatible with economic productivity. Arthritis, for example, owing to diverse etiologies, in Roberts's time often precluded gainful employment and limited consumption of other than medical goods and services. Thanks to medical innovation, joint disease is far more manageable and consistent with an active lifestyle conducive to material production and consumption.

Many other examples exist, documenting a positive impact of medical innovation on formerly chronic debilitating diseases. This fact, plus the declining birthrate in developed countries, implies that on balance medical innovation, and its effect on longevity, is economically beneficial. Breakthroughs in the management of unmet medical needs, such as dementia, will confer even greater value.[59]

If, as argued here, the government-academic-biomedical complex is a relatively poor system for medical innovation, why has it lasted so long and received so little critical scrutiny? The retrospective analysis provided in this chapter provides unpredictable—and predictable—reasons.

Unpredictable, in part, was how the system changed. At the GABC's onset, medical research took place in a few institutions such as New York's Rockefeller Institute or in not-for-profit hospitals affiliated with universities. The hospitals were nonprofit because little profit was possible. The relatively limited availability of medical technology relegated clinical care predominantly to observing hospitalized patients whose illnesses did or did not resolve spontaneously or responded to limited available treatments. Therefore the physicians staffing these hospitals had abundant free time to pursue research in their laboratories, often making measurements on blood or other body fluids taken from patients. Thanks to this leisure and the prestige associated with research, clinician-researchers were willing to sacrifice some income for providing clinical services. The lack of reimbursement restraints at that time still enabled these individuals to sustain comfortable incomes. The bounty of insights yielded by the first medical innovation era meant that researchers working under these relatively leisurely circumstances could only perceive an influx of federal research funding as a great opportunity to amplify their efforts and increase their chances of making discoveries that would advance their careers. The incipient GABC amply supported that perception.

But the system changed. The very success of medical innovation transformed university hospitals into intensive service providers, drove up costs, and accrued an expensive bureaucratic apparatus to manage the service business. Hospital survival, dependent on financial success in providing services, put research on the back burner. One aspect of the bureaucracy—warmly embraced by the GABC—was ever-increasing regulatory oversight, such as over the ethics of experimentation with humans and animals, over the performance of laboratory tests, and over researchers' interactions with industry. These regulations addressed some real but often theoretical concerns and have impeded research.

Predictable, on the other hand, has been the institutional response to these systemic changes based on the fundamentals of human nature and on a consequence of the history of medicine itself.

The human element is that the bureaucracy created by the massive corporate expansion of the biomedical enterprise has created a managerial class with few incentives to challenge the system—and strong incentives to suppress dissent, because disruptions threaten managerial security. Most researchers have no financial or political wherewithal to express dissent. Unlike most business leaders, medical managers have relatively little accountability because the metrics of success in healthcare are so vague and unpredictable. The few investigators who have played the GABC successfully determine what research topics are fashionable and thereby exert regulatory capture of the granting and publication system based on peer review.

Even the leadership of the medical products industry has little motivation to challenge the GABC. It may not enjoy the rhetorical attacks of pharmaphobia, but absent price controls—something its lobbyists have managed to prevent so far—they represent no serious challenge to the industry's profitability. The industry's investors don't care where the science resulting in profitable products originates.

The esoteric nature of science in general and medical innovation in particular has helped to accommodate the widely held conceit that government-supported academic research is the source of most medical innovation. Enactment of the reform recommendations summarized here will require public education concerning the fallacy of that view. Arguably, those who ought to be most motivated to promote those reforms are individuals who personally suffer from diseases that are currently poorly managed by available medical technologies or who have friends or relatives in that circumstance.

REGULATION OF CARCINOGENS AND CHEMICALS
What Went Wrong

s a professor of toxicology in the School of Public Health and
Health Sciences at the University of Massachusetts, I teach a
course called Environmental Risk Assessment. Among the topics covered are the history of environmental legislation, how such
legislation created the legal framework to establish environmental exposure standards, and the scientific basis for regulation of chemical
carcinogens and ionizing radiation. In the course we also place these
types of regulation in their social, political, economic, and international contexts.

Current regulations are based on a deliberate misrepresentation of
the scientific basis for the dose response for ionizing radiation-induced
mutations by the former leaders of the radiation genetics community.
In 1956, these actions culminated in a successful attempt by the prestigious U.S. National Academy of Sciences (NAS) Biological Effects of
Atomic Radiation Committee (BEAR I)-Genetics Panel to manipulate
the world's scientific community and the general public when it recommended that the dose response for radiation-induced mutation be
changed from a threshold to a linear dose response.

FIGURE 7.1
Comparison of the three leading toxicologically based dose-response
models (threshold, linear, and hormetic) used in risk assessment

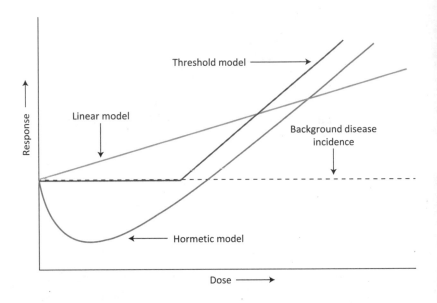

This seemingly minor technical matter can best be shown with a
simple illustration (see Figure 7.1). This model, called the linear no-
threshold (LNT), searches for the lowest exposure to a carcinogen or
ionizing radiation that is associated either with substantial mutations
or with cancer itself. That should be the starting point for any dose-
response model, but it is not. Instead, a line is drawn backward from
the detection threshold data point to the origin on the graph. The implication is obvious. By forcing the response through the origin, any
exposure—including the most minuscule—is claimed to be dangerous.

The National Academy's recommendation to switch to a linear dose
response misled the world's regulatory community, affecting carcinogen standards for chemicals and ionizing radiation to the present day.
A small group of ideologically oriented and prestigious scientists took
advantage of their status and opportunities to institutionalize a demonstrably wrong paradigm concerning carcinogen regulation. While
they no doubt thought their actions virtuous, they were also willing to

blatantly disregard scientific norms in support of the paradigm. Their revolution has been successful for more than half a century—that is, through more than two generations of scientists and the entire modern regulatory history, predating the Environmental Protection Agency (EPA), Occupational Safety and Health Administration (OSHA), National Institute for Occupational Safety and Health (NIOSH), Agency for Toxic Substances and Disease Registry (ATSDR), and influential professional societies, including the Society of Toxicology (SOT), the Society for Risk Analysis (SRA), and many others. The science behind this revolution was known to be false at the time it was undertaken. The LNT model became accepted as truth and was uncritically incorporated into environmental toxicology and risk assessment history, becoming unquestioned dogma and a core scientific belief. It is passed on to new generations of toxicologists, regulators, policymakers, and the general public as part of our scientific and regulatory culture.

Although the situation described above is bad enough for the LNT, the regulatory community also adopted the threshold dose-response model for noncarcinogens, assuming that it would make accurate predictions of responses in the low-dose zone. However, no person or group within the entire regulatory edifice of multiple federal and state agencies in any country ever validated the capacity of this model to make accurate predictions in the low-dose zone. Regulatory actions would be highly dependent on this model to protect the public health. Some 70 years later, when vetting of the regulatory dose-response models finally took place, the threshold dose-response model (as well as the linearity dose-response model) was found to be an abject failure.[1] The next section of this chapter supplies the supporting facts.

A further issue is that global regulatory agencies, as well as many in the scientific community, are not aware that the risk-assessment regulatory framework is built on a series of misrepresentations, unexamined assumptions, and erroneous beliefs that were later seen as complex social constructs, including an intriguing combination of self-interest and transgenerational inheritance of unquestioned concepts. Regulatory scientists and administrators benefited from the existing LNT paradigm—so much so that they didn't do the simple falsification and validation experiments that are at the core of science.

In the first decade of the 21st century, I became interested in determining how the scientific and regulatory communities had validated the threshold dose-response model. After several months of actively searching for articles and any related material concerning the validation of the threshold dose-response model, I began to wonder if it had ever been done. I then called a number of senior toxicologists both in and outside of government asking if they could direct me toward the validation studies. None could. I then renewed my efforts, trying new search strategies to uncover any validation studies, yet none were successful. I then decided to include mention of these failed attempts in seminars and other presentations, hoping to inspire someone in these scientific audiences to come forward with information on such validation. No one ever did. At some point, approaching two years into my search, I came to the conclusion that the threshold dose-response model had probably never been validated during the entire 20th century. This led me to conclude further that entire national standard-setting programs were built on an assumption rather than a reality. It was a paradigm of scientific correctness that had been passed down from one generation of scientists and regulators to another, codified into textbooks, legislation, and hazard assessment protocols for all chemicals and drugs. It is truly mind-boggling that no one ever challenged it.

Having never discovered any prior effort to validate the threshold dose-response model, my colleagues and I set out to do so with multiple independent data sets. Our broader intention was to test the capacity of the threshold, linear, and biphasic models to make accurate predictions in the low-dose area.[2] These efforts, which resulted in the publication of several papers in leading toxicology journals, revealed that both the threshold and linear dose-response models consistently and uniformly poorly predicted responses in the low-dose zone.[3] In fact, their respective performances were far below any fair-minded acceptable level. The only dose-response model that performed consistently very well in each of the evaluations was the biphasic one. This is not surprising since it forms the basis for modern pharmacotherapeutics.

The findings revealed that the threshold dose response, which became the only model of choice by the medical, scientific, and regulatory communities starting in the late 1920s for chemicals and radiation exposures, was shown to have very significant limitations. In addition, my concurrent research on the biphasic dose response had already revealed many thousands of cases in which the threshold dose-response model also failed to predict responses in the low-dose zone.[4] We found that the entire regulatory edifice for chemicals and radiation had been built on a failure of the scientific, medical, and regulatory communities to ever take the time to vet the model on which they were betting the health of the human population.

The similarity between the fiascos of government dietary advice and of regulatory science is painfully obvious. One caused an epidemic of obesity and type 2 diabetes, and the other caused countless morbidity by "outlawing" things that would be beneficial in small doses. This is diametrically opposed to the therapeutic model and the hormetic response, and it shows the lengths scientists will go to in order to protect an established, but failing, paradigm.

Assuming that one intellectual generation lasts about 20 years, then we have now passed through four to five generations of physicians and scientists who accept the threshold model on faith. A serious problem with this situation is that the scientific, medical, and regulatory communities had every professional incentive to protect the existing paradigm. They marginalized the biphasic dose response by continuing to associate it with the high-dilutionist wing of the homeopathy party, an easily ridiculed group. The strategy worked, even if it was based on a falsehood.[5] However, oft-repeated falsehoods become truths. Biphasic dose responses were not seen often, and when they were seen they were trivialized and discounted. We are still in the early stages of Thomas Kuhn's model for paradigmatic change:

In science . . . novelty emerges only with difficulty, manifested by resistance, against a background provided by expectations. Initially, only the anticipated and usual are experienced even under circumstances where the anomaly is later to be observed.[6]

We are now at the stage where only a very few are "observing the anomaly," in no small part because there is very little incentive and a lot of professional downside in doing so. When this lasts for multiple scientific generations, an alternative, more explanatory paradigm is systematically suppressed.

Linear No-Threshold: Origin and Implications

Ultimately the threshold model for ionizing radiation would morph into the LNT. This development was an outgrowth of the 1927 research findings of Hermann J. Muller that very high doses of x-rays could cause gene mutations in the mature spermatozoa of the male fruit fly, a discovery that took nearly two decades of intense focus and much competition.[7]

The first challenge to the threshold concept by the mainstream scientific community was offered by two physical chemists from the University of California, Berkeley, one of whom was the internationally famous Gilbert Lewis, who would be nominated for the Nobel Prize some 42 times before dying in 1946 from cyanide poisoning in his laboratory. They were seeking a mechanistic understanding of the process of evolution, one of the most significant questions of that time period.[8]

Following the discoveries of Charles Darwin and Gregor Mendel, there was great interest in trying to discover the mechanism by which evolution occurred. It was believed that evolutionary change must be mediated through gene mutation. The problem was that, since about 1910 and for the next 17 years, no one had been successful in inducing mutations using a whole host of toxic agents and different types of radiation. With Muller's breakthrough, Lewis proposed that the mechanism of evolution was gene mutation caused by cosmic rays and background terrestrial radiation. Because the dose is so small, he assumed that the dose response would have to be linear at a low dose. As a result, Lewis's hypothesis was only able to account for about 1,300th of the background mutation rate, based on Muller's fruit fly data and assuming a linear dose-response relationship.[9] Muller retained his commitment to understanding the causes of evolution, but it seemed pretty clear that the mechanism would not be found in Gilbert's background radiation hypothesis.

Although Muller did not support Lewis's arguments that cosmic and terrestrial radiation were the driving force for evolution, he soon directed several students to assess the dose-response features of the x-ray treatment. Muller ultimately won a Nobel Prize for this, but his research in fact did not yield a linear dose response. Of his three experiments, the third lacked a control group, thereby preventing a firm assessment; the second gave the suggestion that the responses varied nonlinearly with dose but with the square root of the dose, while the initial experiment yielded data consistent with a threshold dose response.[10]

His students' research continued to employ what amounted to very high doses. Nonetheless, these researchers reported linear dose responses. Based on these findings, Muller developed a strong belief that the dose response was linear and that linearity would extend down to a single photon.[11] The lowest doses tested in their linearity-supporting articles were about 300,000-fold greater than background. Furthermore, Muller was selective in choosing which data to focus on, as there were other contemporary credible findings displaying a threshold perspective.[12] In fact, on balance there was more support for a threshold dose-response model interpretation than for a linear perspective. Yet Muller would give the impression that he passionately and firmly believed that the dose response for gene mutation was linear in the low-dose zone. It is difficult to understand how Muller would make a "commitment" to a dose-response model for low-dose predictions when there was no testing in the low-dose zone!

Despite the use of extremely high doses and other data that challenged a linear perspective, Muller would nonetheless introduce the term "proportionality rule" in 1930 to define the nature of the dose response for ionizing radiation-induced gene mutation in the low-dose zone.[13] He was claiming that ionizing radiation and gene mutation were different from other agents and endpoints, respectively. He claimed that radiation-induced gene mutations do not show a threshold but act in a linear fashion. Soon thereafter, the proportionality rule gained traction within the radiation geneticist community, in large part because Muller had become famous as the first to induce changes in heritable material.

With the discoveries of Muller, the American X-ray and Radium Protection Committee concluded that there was enough evidence for them to be concerned about genetic injuries.[14] Nonetheless, such discussions did not result in policy changes and the threshold model retained its dominance. Based on the recommendations of Stanford's Robert Reid Newell, the 1940 committee did adopt a 50 percent reduction of the radiation exposure standard, down to 0.05 r/day.

The Manhattan Project and the LNT

The Roosevelt administration created the Manhattan Project in response to Albert Einstein's 1939 letter to President Roosevelt about the likelihood of a German effort to build an atomic bomb (see also Chapter 1). As part of this effort, research was undertaken to better assess the nature of the dose response to ionizing radiation. The key research on mutation was to take place at the University of Rochester under the direction of Dr. Donald Charles for mouse studies and Dr. Curt Stern for fruit fly investigations.

In the case of the mouse work, no detailed papers were ever published. Only two brief summaries appeared, one in 1950 and the other in 1961. It is not clear why the mouse work did not yield the type of productivity and insights expected, despite the use of nearly 400,000 animals. Regrettably, life did not end well for Charles; he left his position at Rochester and, in 1955, committed suicide in a Manhattan hotel, leaving a despondent letter reflecting both job and marriage failings.

Stern's research, however, would revolutionize the field of cancer risk assessment and eventually lead to the institutionalization of the LNT. Stern was a towering giant in genetics, having discovered a component of chromosomal crossing-over, a mechanism that is a key foundation of modern genetics. Soon after joining the Manhattan Project he appointed Muller as a formal consultant. Muller and Stern firmly believed that the experiments with the fruit fly would provide strong support for the LNT model, confirming their long-held views. Their efforts were the most extensive to date.

Furthermore, they would be working with highly experienced individuals. In their first major experiment under the direction of Warren Spencer, an Ohio State University PhD graduate and a fruit fly expert on leave from the College of Wooster in Ohio, they found that acute doses of x-rays (i.e., total dose given over 2–40 minutes) appeared to cause a linear dose-response relationship for germ cell mutations. This research, which became highly acclaimed, had many important methodological limitations, especially affecting the validity of low-dose responses. Yet these limitations were either ignored or missed by experts in the field. A detailed internal letter by Muller in 1945 assessing the Spencer research also failed to identify any of the many now-recognized weaknesses of this research.[15]

A follow-up study by Ernst Caspari was designed to assess the effects of gamma rays that were administered throughout the entire life of the fruit fly, providing chronic exposure. The comparative dose rate in the Caspari experiment was about 1/13,000th of the lowest-dose rate given by Spencer. Even though both studies administered the same total dose, Spencer delivered it in a few minutes, while it took three weeks in the Caspari study. According to the LNT model, both studies should have resulted in the same amount of damage since ionizing radiation was assumed to cause damage that was cumulative and irreversible.[16]

It did not. To the surprise of Caspari, the mutational data were not cumulative, revealing an apparent threshold response. In effect, Caspari's data challenged the assumption that mutation damage was best predicted by total dose—a key feature of the LNT model—rather than by dose rate. Stern initially did not accept these findings, claiming that the data were most likely an artifact of the experiment, with the control group displaying a much higher than normal response, thereby leading to the threshold, rather than the linear dose response. Caspari dug in, searched the literature, and found a number of papers that supported his position that the control group had responded normally, forcing Stern to back down, at least temporarily.

While Stern was forced to retreat on the issue of the control group, he intervened with a more powerful strategy that had the same effect, which was to not accept the threshold dose response. He forced the discussion of the Caspari paper to conclude that it was not possible to accept the threshold interpretation until it was determined why this study obtained different results than the earlier Spencer work. Stern knew that the two studies had at least 25 significant methodological differences, making them virtually impossible to properly compare. For example, the studies differed in the type of radiation used, the diets the flies were reared on, which sex was exposed to the radiation, the temperature of the study, and many other factors. To figure out exactly why the two studies obtained different results would be a major undertaking by itself, a task that was never attempted by any of the original researchers, any of Stern's or Muller's subsequent students, or anyone else since.[17]

Stern sent this draft manuscript to Muller for review just before traveling to Stockholm to receive the Nobel Prize in December 1946. On November 12, 1946, Muller wrote to Stern telling him that he was shocked by the findings, that the data offered a strong challenge to the LNT concept, and that the studies needed to be replicated as soon as possible. Yet he could not criticize Caspari, for he was a competent investigator.

Muller traveled to Sweden and received his honor. Despite just having read Stern's paper, in his Nobel Prize lecture he stated that it was no longer possible to even consider the possibility of a threshold dose response for ionizing radiation. The risk assessment field had to switch from using the threshold dose-response model to the LNT model. He made these comments after having just seen the most extensive set of data on the topic, from a research team that was considered very experienced, where Muller himself provided his own Muller-5 fruit fly strain and was a very involved consultant. Detailed letters between Muller and the Rochester team document the extensive role that Muller had in helping to shape their research strategies, study design, and research methods.[18]

Within several weeks of delivering the Nobel Prize lecture, Muller would restate the need to replicate the troubling threshold-supporting study, that he had no technical criticisms of Caspari's research, and that his paper could be published given all the caveats he insisted upon. As a result, Caspari's result should not have been considered valid until it was determined why his findings differed from those of the Spencer experiment.[19]

Muller's Nobel lecture was historic and its worldwide influence profound. Proclaimed in the shadow of Hiroshima and Nagasaki, Muller's concerns about the effects of radiation on the human genome were a major jolt. The problem was that Muller knew better. He knew that the strongest and most relevant study to date (i.e., the Caspari chronic study) actually supported the threshold model. The story gets even more interesting after that.

Muller and Stern: How Their Omissions Shaped Cancer Risk Assessment

Back in Rochester, Stern sought to replicate the research of Caspari. With the end of the war, Caspari moved on to Wesleyan University in Connecticut and Spencer returned to the College of Wooster.

Stern had a new graduate student named Delta Uphoff. In her first major experiment she and Stern found a serious problem. Her control group mutation rate was about 40 percent lower than what had been reported in the published literature and the Caspari study. In their formal manuscript submitted to the new Atomic Energy Commission (AEC) they dismissed their data as uninterpretable, blaming the results on investigator bias. This was an extraordinary statement that was not discussed further, beyond acknowledging the classified paper. Was the bias that of Uphoff or Stern or both?

These results were very troubling to Stern, who then engaged Muller in a series of letters to resolve the control group issue. This discussion took place over a six- or seven-month period in the first half of 1947. He had already "lost" this debate with Caspari when the published literature was considered. However, Stern knew that Muller was sitting on top of a very large database of control group findings. In these letter exchanges Stern received information that the control group data of Muller were strongly supportive of the Caspari research. Furthermore, Muller had tested control group aging mutational responses in a variety of fruit fly strains, and the data were very consistent.

For unknown reasons, Uphoff's findings were far off the mark, making them appear unreliable. Muller continued to study this critical issue after he arrived at the University of Indiana, publishing multiple papers with a series of students and all with similar data. Without question, the findings of Caspari were strengthened while those of Uphoff remained in the uninterpretable zone, or so it was thought.

Uphoff's next experiment also uncovered an anomalously low control group mutation rate. A third and final experiment led to yet a new and troubling irregularity—that of a treatment effect severalfold greater than an estimated linear response.

What to make of these data was the challenge to Stern. In the end, it appeared that his research would suffer the same fate as Charles's mouse research. After all the effort, the findings were not yielding the expected support for the linear dose-response model. Furthermore, the data from Uphoff's three experiments were unreliable.

As the pressure mounted on Stern, he opted for a new strategy that would change the face of environmental risk assessment for the next half-century and beyond. He decided to revert to his original position and discount the findings of Caspari. It was Stern and Muller's best hope to save the linear single-hit dose response. However, he also added a few more academic wrinkles. He then also reversed his position on the findings of Uphoff, claiming that what were "uninterpretable" less than a year earlier were now fully valid and acceptable. These judgments were without any apparent scientific basis, with no explanation provided.

In fact, Stern did not inform the scientific community that he had changed his position from his earlier critique of the Uphoff data. Stern's reinterpretation was hidden in a highly classified manuscript sent to the AEC that he knew would be seen by only a very few. In the paper's acknowledgments, Stern and Uphoff stated that Muller had approved their use of his data, which supported the validity of the Caspari control, and that the control group results of the Uphoff study were aberrantly low. This acknowledgment is of historical importance because it connects Muller to the Stern and Uphoff manuscript and his support to the Caspari control group. In fact, letters exchanged between Stern and Muller do this as well. Muller knew that Caspari's control group mutation rate was normal and that his data were used to support this determination. This point will become important later when Muller bizarrely, publicly, and repeatedly claimed that the control group mutation rate in the Caspari experiment was aberrantly high. Stern also failed to acknowledge the numerous methodological limitations of the Spencer paper, such as irregular control of temperature, the inappropriate combining of various treatment groups especially in the critical low-dose zone, and the inconsistent standardizing of the x-ray equipment, among others.[20]

When Stern did publish his findings, all five experiments (i.e., the Spencer, Caspari, and Uphoff research) were summarized in a table in a one-page technical note in the journal *Science*.[21] Since many data were missing and no methods were presented, he stated that a follow-up detailed paper would be published with all the methods and data. However, in a significant failing, Stern never published the promised follow-up paper.

It is more than curious that *Science* would have published such a note without any supportive material, suggesting that Stern may have used a personal contact at *Science* to get his paper published. Several months before Uphoff and Stern submitted their manuscript, Bentley Glass, Muller's first PhD student at the University of Texas, who had received a Fulbright Fellowship to work with Stern, became an editor at *Science*. Given the expertise of Glass in fruit fly mutation, it is hard to imagine that he was not involved in some oversight of the Uphoff and Stern manuscript.[22]

This was not the first time that *Science* would give research related to Muller a strong push forward. For example, when Muller published his Nobel Prize–securing paper in *Science*, it also did not contain any data. All his paper provided was a detailed discussion of the findings. How was this possible? It seems that Muller knew he was in a race to be the first to show that mutations could be artificially induced. The only way he could win the race was to put the cart before the horse and present only the discussion. Without such deferential treatment from *Science*, he might have come in second, since others such as Lewis Stadler at the University of Missouri were not far behind.

In the years after *Science* published Uphoff and Stern's technical note, the issue of a Caspari and Stern study indicating chronic low-level radiation effects surfaced repeatedly.[23] However, Muller was quick to assert that the findings were problematic since the control group mutation rate was uncharacteristically high and this was the principal reason why the Caspari and Stern data showed a threshold.[24] While Muller offered such comments in the published literature, his own data and various documents discussing his data confirmed that he repeatedly and strikingly contradicted himself.[25] This behavior was quite risky since the contradictions would have been easy to show if one were willing to confront this scientific icon and his group of loyal radiation geneticists. His contradictions were necessary to support the LNT model, and they were also needed to support his remarks in the Nobel Prize lecture. However, no one stepped forward to confront Muller or Caspari.

By the early 1950s the radiation genetics community had been won over: it accepted the LNT model and rejected the threshold model for mutation. This talented group had never investigated in adequate detail the Spencer, Caspari, and Uphoff papers that were coauthored with Stern. The numerous critical limitations of the Spencer paper never surfaced. Likewise, the "uninterpretable" findings conclusion concerning the Uphoff data remained unknown. The only critical focus was directed toward the Caspari study, and this was due to the false charge of a high control group mutation rate as perpetuated by Muller and Stern.

With the acceptance of the LNT model, the regulatory agencies uniformly appealed to the authority of Muller and Stern, the leading radiation geneticists, accepting their "uniquely informed insights" on the most influential studies. Confirmation bias is common in science and in life, but in this field it provided a type of preconditioning for the perfect risk assessment storm.

At a critical time in environmental public health history, the National Academy of Sciences convened a major assessment of the public health implications of low doses of radiation, including their mutational effects. Concerns about public exposure to ionizing radiation rose during the early 1950s, when the United States and the Soviet Union were in an arms race and were actively conducting above-ground testing of their atomic weapons. Despite their highly attenuated concentrations, radionuclides quickly achieved worldwide distribution, finding their way into foods and drinking water. Exposure to ionizing radiation was being rapidly expanded from patients and medical personnel to the general public.

Until this time, Muller and his group of radiation geneticists had always been outvoted or outmaneuvered on the issue of the nature of the dose response—that is, the threshold versus linearity debate. Now the genetics panel of the NAS committee had the votes to switch the risk assessment paradigm. Finally, they had their chance to protect the human genome from radiation-induced mutation and the occurrence of heritable diseases as well as cancers. In fact, the radiation geneticist community of that era deeply believed that it was their responsibility to save future generations from genetic harm. No other discipline had their knowledge, experience, and unique insights. This was not only their time, but also their opportunity to change risk assessment policy in the United States and worldwide.

For two decades the radiation genetics community had fought to change the risk assessment paradigm for ionizing radiation. They wanted it based on heritable changes, and they wanted the dominating threshold model that had pervaded the medical community replaced with the LNT model, which was born from Muller's research with mature fruit fly spermatozoa and transformed into a regulatory mechanism based on the assumption that a single hit could induce a mutation, which could possibly lead to birth defects or cancer in humans.

The National Academy of Sciences BEAR I Genetics Panel: How It Misled the World on Risk Assessment

One might have thought that the historic convening of the NAS Genetics Panel in 1955–1956 would produce a major debate over whether the LNT model should replace the threshold dose response for assessing the risk of mutation and cancer. It would be a test of the wills, the historically powerful medical community versus the upstart radiation geneticists led by their Nobel Prize winner, Hermann Muller. At least that is how I expected to find that the series of meetings of these two groups of scientific and medical leaders would go. I obtained the transcripts of their meetings, eager to read the debate and to learn what arguments had been the most persuasive and who had led the way.

To my surprise, there was no debate at all. In fact, it became quite clear that a decision had been made before the convening of the Genetics Panel that LNT was in and threshold was out, an observation later confirmed in the historical writings of the unusually named scientist Dr. Jim Crow, the last member of the Genetics Panel to die.[26] All that was needed was to stack the panel with the correct people and the votes would be there.[27] The Genetics Panel was highly leveraged to support the Muller perspective. It wasted no time debating the merits of LNT and threshold. The group simply adopted an LNT perspective and then moved on to other issues.

During their assessment, the BEAR I Genetics Panel falsified and fabricated the research record concerning the estimate of radiation-induced genomic risk, as I have documented in detail.[28] It made a decision not to share the profound degree of uncertainty among the panel members, but rather to misrepresent it by removing and changing data concerning estimates of genetic mutations in the U.S. population at a certain level of radiation exposure.

The panel members were worried that the scientific community, government officials, and general public would not accept their radiation health/risk policy recommendations, including the adoption of LNT for risk assessment, if they were honest and shared the sizable uncertainties. They deliberately misrepresented the research record in their formal panel publication in the journal *Science*. When a group of prominent biologists requested that the panel provide the documentation for their policy recommendations, they voted not to, a vote that was shared with the president of the NAS, Detlev Bronk, thus linking the deceptive practice right to the top. All these actions of the panel are documented in the historical record (see endnote 24).

This set of coordinated deceptions was used to persuade other scientific bodies and the international community to follow their leadership and direction. Yet this course of action was surprisingly easy. It started with the activism and leadership of the Rockefeller Foundation, which provided the funding for the panel in the first place, printing many thousands of copies of the deceptive report and distributing it widely, including to all public libraries in the country. It was also soon arranged to have the journal *Science* lend its immense credibility to the dose-response switch by permitting the panel to publish a substantial paper on their conclusions.[29] Congressional hearings were also convened, and members of the panel testified, advocating the use of the LNT and basing it largely on the publications of Stern and his colleagues,

Spencer, and Uphoff. *Science* would play a further role in the process by publishing a profoundly influential paper by the future Nobel Prize winner Edward B. Lewis on cancer and radiation exposure, a paper that was heavily dependent on the Uphoff and Stern technical note.[30] This article obtained a further boost from *Science* once again, receiving a ringing endorsement in an accompanying statement by the journal's editor-in-chief.[31]

As noted in an oral history, Lauriston Taylor, chairman of the National Committee for the Radiation Protection and Measurement (NCRPM), stated that other national and international radiation committees, including his NCRPM, were quietly waiting for the proclamations of the NAS panel, as it carried substantial authority. Once the NAS report was made public, these committees soon recommended the switch to the LNT and its generalization to somatic cells so that the LNT could be applied not only to mutational events, but also to cancer risk assessment. Further speeding up the process was the fact that some members of the NAS Genetics Panel served on several of those other panels, thus extending their impact and influence. In effect, these radiation scientists got to vote two or three times, further stacking the deck and getting multiple bites at the apple.

Within a very short time, the radiation geneticists had led a profound environmental and medical policy revolution. They got the U.S. government and the world regulatory community to adopt a new risk assessment paradigm, even if it was a little short on real data.

There was now a global consensus that there was no safe exposure to ionizing radiation, that the dose response was linear. Even if one could not measure the impact scientifically, it was assumed that even a single ionization would cause permanent damage and increase the risk of genetic damage and cancer.

From 1955 to the end of the decade, the most significant changes in environmental policy occurred. Not long after the publicity associated with the linearity recommendation of the BEAR I Genetics Panel, its impact began to be seen in the scientific literature: R. R. Newell applied LNT to population-based mutation frequency; and C. Buck, A. Grendon, and others applied the LNT concept to cancer risk assessment.[32] This panel's actions also profoundly ramped up fear of the continuing arms race and probably influenced the United States and the Soviet Union to end above-ground testing. The adoption of the LNT was also at the heart of the creation of something called the "precautionary principle," a highly conservative way of thinking about risks from emerging technologies, chemicals, and radiation.

There was a major problem with all the success of the BEAR I committee, and that of the radiation geneticists. The science on which they based their good intentions was known by the leadership to be in error. In retrospect, the Genetics Panel of the BEAR I Committee failed to vet the principal research findings on which their recommendation to switch from a threshold dose-response model to the LNT was based. The committee had simply come under the leadership spell of Muller and his longtime and unrelenting advocacy of the LNT. For Muller, his 25-year crusade was finally realized. From the time he first introduced the proportionality rule into the literature in 1930 until it was realized in the recommendations of the BEAR I Committee Genetics Panel, he had been the champion of the LNT movement. Muller would do what he believed needed to be done to achieve this goal.

Muller would also do his best to prevent the publication of papers that might erode belief in the LNT, as he attempted to do in the case of MIT professor Robley Evans (1949), who rejected the LNT.[33]

Throughout it all, there is very little evidence that Muller's peers confronted him on any of these key issues. However, there was an instance when Muller was so upset that a paper in *Advances in Genetics* had been published without his having first reviewed it that he threatened to resign from the editorial board, berating the author and the judgments of the reviewers and the editor. So offensive were the written comments of Muller that Milislav Demerec, the journal editor, did stand up to Muller; he defended the journal and the review process and told Muller that he was dangerously close to imposing his version of censorship on the journal.[34] With the line drawn in the sand by the feisty Demerec, the powerful Nobel Prize winner struck back by resigning from the journal, a move that insulted Demerec, who had a long record of helping to ensure that Muller's research would be funded.

Despite his willingness to stand up to Muller, Demerec was just as committed to the LNT as Muller, Stern, and the others. In fact, when

Caspari shared his threshold-supporting findings with Demerec in the fall of 1946, a disturbed Demerec challenged Caspari with the same question that was also being considered by Stern and Muller: "What can we do to save the hit model?" It was not about following the data, but rather about preserving a dogma and a paradigm cloaked in virtue and virtue-signaling—that is, protecting the public from certain harm.

While it may be possible to see this group of radiation geneticists as misguided in their romantic heroism to save the world from genetic harm caused by nuclear testing and the excessive use of medical x-rays, information is also available in the preserved private correspondence of some of the members of the BEAR I Genetics Panel to suggest that professional self-interest was a commingled prime motivator in their fight to "ban the bomb." The majority of the Genetics Panel were active researchers in university settings. Just like today, there was a publish-or-perish mentality. Research dollars were critical, and such resources were provided by governmental organizations and foundations for these investigators. Almost all of the academic members of the Genetics Panel were externally funded. While competition for funding was therefore central to the process, letters indicated that some members of the BEAR I Genetics Panel also deliberately overstated the nature of the radiation dangers to the public health and used highly inflammatory language to make their research seem more frightening and important—that is, more fundable. By advocating strongly for the LNT, the members of the Genetics Panel would make their area of paramount importance, enhance their professional opportunities to advise the government, receive research funding, and engage in consulting activities.

Academics and Self-Interest Science

In his 2007 dissertation, Michael Seltzer provided evidence that members of the Genetics Panel clearly saw their role in the NAS committee as a vehicle to advocate and lobby for funding for radiation genetics. Moreover, it was hoped that the committee, which continued to exist for many years, would influence the direction and priorities for future research funding.[35]

Such hoped-for funding possibilities for radiation geneticists can be seen in correspondence between Chicago's George Beadle (also a Nobel laureate), Columbia's Theodosius Dobzhansky, Muller, and Demerec. For example, Demerec, the head of genetics research at Cold Spring Harbor Laboratory, offered a possible funding plan that

> could be accomplished by setting aside a fund, (let us say, one hundred million dollars) to be administered by some competent organization (such as the National Academy of Sciences) and used during a period of twenty or twenty-five years to fund already functioning research centers so as to attract and train first rate scientists.

Dobzhansky reacted to Demerec's proposal by indicating that he would,

> needless to say, be all in favor (of) $100,000,000 for research in general genetics. . . . But I would find it hard to keep a straight face arguing that they [general genetics] must be studied to evaluate the genetic effects of radiations on human populations.[36]

Demerec responded:

> I, myself, have a hard time keeping a straight face when there is talk about genetic deaths and the tremendous dangers of irradiation. I know that a number of very prominent geneticists, and people whose opinions you value highly, agree with me.[37]

Dobzhansky responded:

> Let us be honest with ourselves—we are both interested in genetics research, and for the sake of it, we are willing to stretch a point when necessary. But let us not stretch it to the breaking point! Overstatements are sometimes dangerous since they result in their opposites when they approach the levels of absurdity. Now, the business of genetic effects of atomic energy has produced a public scare, and a consequent interest in and recognition of (the) importance of genetics. This is to the good, since it will make some people read up on genetics who not have done so

otherwise, and it may lead to the powers-that-be giving money for genetic research which they would not have given otherwise.

Seltzer concluded that these letter exchanges demonstrated that the geneticists were very concerned about the viability of their profession and that they clearly recognized their "moment" in the government and public spotlight. They definitely wanted to find a way to manipulate the current situation to improve their opportunities for increased funding.

In their letters, the geneticists revealed that their scientific and social goals were also entangled with their need to advance their research area and career success. It is striking to read phrases like "finding it hard to keep a straight face" and a willingness to "stretch a point" all in the name of self-interest. Both phrases indicate a capacity and willingness to distort the truth in the name of self-interest. Self-interest may have played a role in their failure to vet Stern's research findings and in their failure to demand that Stern provide a detailed paper after publishing the note in *Science*. These types of biases are critical, for they can affect the outcome of a committee's recommendations. The situation with the BEAR I Committee Genetics Panel was clearly problematic on multiple levels.

Muller, BEAR I, and the EPA Cancer Risk Assessment: The Connection

The recommendations of the BEAR I Genetics Panel proved to be of profound significance for the cancer risk assessment of ionizing radiation.[38] The recommendation of the panel had a long reach, across decades, agencies, and agents. Although intended for ionizing radiation, the LNT concept could be generalized, and once it could be reliably shown that chemical carcinogens were often mutagens it would not take too long for the newly emerging EPA to make the BEAR I dose-response recommendation its own.

In his history of carcinogen assessment, Roy Albert, the chairman of the EPA Cancer Assessment Group (CAG), wrote that during the 1970s the EPA adopted the LNT of the Atomic Energy Commission that had been applied to exposures from atomic weapon test fallout. He noted that the LNT model was simple and that its simplicity made it attractive to the EPA. In fact, to determine the potential incidence of cancer, all that was needed was to identify the lowest dose of an agent that induced a statistically significant response and draw a straight line to the origin of the graph. The biological plausibility was based on the linearity of the mutation dose response as set within the framework of target theory. Importantly, Albert noted that "any difference between chemical carcinogens and ionizing radiation could be ignored as they both caused genetic damage."[39]

The actions of the EPA to adopt the recommendations of the BEAR I Genetics Panel for its cancer risk assessment activities reveals that the foundation of modern cancer risk assessment in the United States and in most countries was based on incorrect findings.

Of considerable importance is that both Muller's research and that of Stern during the Manhattan Project period were performed with mature spermatozoa. Muller, Stern, the Drosophila-dominated radiation genetics community, and advisory committees around the world extrapolated findings from mature spermatozoa to somatic cells, not realizing that mature spermatozoa lacked the capacity to repair damaged DNA. It was this inability to repair DNA within the mature spermatozoa that led to its heightened mutational susceptibility and its capacity to grossly overestimate mutational risks in other reproductive and somatic cells.

This was especially true at low dose rates, and created a belief among the radiation geneticists in the correctness of the LNT model. This belief, which was initiated by the BEAR I Committee Genetics Panel and soon generalized to cancer risk assessment by other advisory groups, was a flaw of major proportions. This action was something like using mutating bacteria lacking DNA repair to estimate genetic risks in humans. Not only did this group of radiation geneticists ignore evidence that conflicted with their policy desires; they also got the science wrong.

The concept of dose rate in radiation would be unequivocally reported as early as 1958 by William Russell and his colleagues.[40] DNA repair would be discovered by Tomas Lindahl in the early 1970s; a given cell repairs itself 200 times a day.[41] Together, these two findings should have been fatal to the LNT and the massive regulatory artifice that it spawned. The massive global community will take a long time to change, especially since it has largely adopted the precautionary principle, making it more difficult to self-correct as new scientific understanding emerges.

Final Thoughts

The paradigm guiding risk assessment for carcinogens and noncarcinogens is based on a demonstrable falsehood. The Environmental Protection Agency uncritically accepted the LNT for the assessment of risks from carcinogens like ionizing radiation and certain chemicals, as well as for toxic chemicals. The agency built its entire regulatory edifice on models that have been shown to exaggerate the risks of substances in the low-dose zone, the location where we spend the vast majority of our time.

It is hard to imagine that the regulatory agencies and all the regulated industries and their subsidiary consultants never thought to examine the foundations on which all the regulations that they debate, fight, and litigate are based. The failure of the scientific community and the regulatory agencies to uncover the unscientific origins of the LNT and how it came to influence the risk assessment process is important because it profoundly affected the development of environmental legislation, health standards in air, water, food, soil, and the health of the population. This failure devolves to the economy, the proper use of natural resources, and the course of environmental and public health history. Correcting this gargantuan error remains a fundamental challenge to, and a necessity for, the regulatory community.

Acknowledgment

Long-term research activities in the area of dose response have been supported by awards from the U.S. Air Force (FA9550-13-1-0047) and ExxonMobil Foundation over a number of years. The U.S. government is authorized to reproduce and distribute for governmental purposes notwithstanding any copyright notation thereon. The views and conclusions contained herein are those of the author and should not be interpreted as necessarily representing policies or endorsements, either expressed or implied. Sponsors had no involvement in study design, collection, analysis, interpretation, writing, or decision to submit.

RADIATION POISONING

mericans are one of the few peoples in the world who, as land-
owners, own both the land and what is beneath the surface. If
there happens to be a rich vein of gold beneath your feet, it's yours
to have, hoard, or sell.

Consider the case of Walter Coles Sr. Under his property, at Coles
Hill, Virginia, in poverty-throttled Pittsylvania County hard up against
the North Carolina border, around 200 miles inland from the Atlantic
Ocean, is the largest known uranium deposit in the United States, and
the seventh largest on earth.[1] The value has been estimated at around
$7 billion.[2]

In late March 1979, a geologist for Marline Uranium was driving a
Hertz rental car in which he had placed a scintillometer on the dash-
board. Much like a Geiger counter, the device detects radioactive de-
cay as it happens. Most of the time, such a machine produces sporadic
chirps. But near Coles Hill the counter went off the chart. The next day,
Henry Singletary, another employee of Marline, verified that the company had found a humungous uranium deposit—one that would be remarkably easy to mine because the deposits were both on and not far
below the surface.[3]

In a remarkable coincidence, that very day, March 28, 1979, a cooling malfunction caused a partial meltdown of a nuclear reactor at Pennsylvania's Three Mile Island power plant. It did not matter to public opinion that the average amount of radiation exposure to the 2 million people who live in the vicinity of the plant was a mere 1/100th of the annual exposure from natural background radiation. What did matter is that a physically impossible accident in a U.S. commercial power reactor was concurrently on the silver screen, acted in and promoted by activist Jane Fonda. *The China Syndrome*, released on March 16, a mere 12 days before the Three Mile Island incident, was a great success even if it was as fictional as a science fiction movie could be. In the movie scenario, a reactor melts down, the container ruptures, and intensely hot uranium or plutonium burns its way down, ostensibly "digging a hole to China."

Three Mile Island and *The China Syndrome* empowered anti-nuclear environmentalists, and the Virginia legislature responded to their concerns by declaring a moratorium on uranium mining, which can only be lifted by another act of the same legislature. It was signed into law by Governor Charles Robb.

In the first decade of the 21st century, because of the upcoming 2013 expiration of a program in which Soviet nuclear bombs were turned into reactor fuel, the price of uranium soared. Instead of selling mining rights and reaping an instant profit, Walter Coles tapped his son, Walter Jr., an energy analyst for a Wall Street hedge fund, to raise enough money to start a new company, Virginia Uranium, which later traded on the Toronto Stock Exchange. Coles soon had raised $22 million to capitalize the new company.[4]

In 2013, the political climate in Virginia was challenging. While the Republican Party held a supermajority in the House of Delegates, and would likely vote to lift the moratorium on uranium mining, the Senate was exactly split, with 20 Republicans and 20 Democrats. Well-funded environmental activists, such as the Southern Environmental Law Center in Charlottesville, and the Piedmont Environmental Council in Warrenton, Virginia, were passionately opposed to uranium mining, and the politics were close and bitter.

As shown in the Introduction, it is easy to understand how the incentive structure of science creates systematic biases in various disciplines. When an extant literature is summarized, the resulting compendium will reflect similar biases. But there is a further distortion that crops up occasionally, which might be called "bias by authority." We will see this again in Chapter 9 of this volume, on Alaska's Pebble mine. This bias occurs when governmental or quasi-governmental entities selectively gather scientific experts in pursuit of a foregone conclusion, and it is the sad story of Virginia Uranium, Inc., which now trades for $0.08 per share as Virginia Energy Resources.

In 1863, President Abraham Lincoln requested and signed legislation creating the National Academy of Sciences to "investigate, examine, experiment, and report upon any subject of science." Of course, what he was really interested in at the time were new technologies and ideas to successfully prosecute the Civil War. Another believer in expansive government, and again in wartime, President Woodrow Wilson created the National Research Council (initially called the National Research Foundation), as the research arm of the National Academy of Sciences.

Congress, the president, or maybe even you—if you have enough money—can ask the Academy to research a topical question and provide scientific advice. The Academy then asks its Research Council to assemble a committee (usually not composed of National Academy members, though some may be included). One way or another, Academy members are almost always a successful product of the Big Science incentive and bias system described in the Introduction and Chapter 2.

Committee members tasked to study a given topic are vetted and well known by the Academy. In controversial areas, there is a majority view balanced by one or a few dissenters. By selecting the participants, the Academy knows the result it will get.

Walter Coles Jr. wanted to play it straight. He proposed that the National Academy commission a report on the state of environmental science and technology with regard to mining his deposit. Surely the Academy would be fair and impartial, he thought.[5] And so, the Academy informed

Coles that indeed the National Research Council would select a committee to do a report, a report, he was informed, that he had to pay for, via a grant administered by Virginia Tech in Blacksburg that was subcontracted to the National Research Council. Committee members do not travel cheaply, and the administrative support, overhead, and production costs added up to the tidy sum that Coles contributed.

Of the fourteen committee members, four were likely to support exploitation because they had either consulted in defense of the industry on environmental matters or been involved in mining themselves. The views of an ecosystem hydrologist and a uranium geochemist on the committee were harder to predict, though the hydrologist was a board member of a local environmental activist organization near his home institution, Frostburg State University in the rugged Maryland panhandle.

The other eight gave clear indications that they were predisposed against exploitation. Many were strict believers in the linear no-threshold (LNT) model, discussed at length by Edward Calabrese in Chapter 7 of this volume. These panelists were clearly strongly risk-averse, and their views were dominant and reflected in the subsequent report, *Uranium Mining in Virginia* (hereafter *UMV*), released on December 19, 2011. The centerpiece of the report concluded:

> If the Commonwealth of Virginia rescinds the existing moratorium on uranium mining, there are steep hurdles to be surmounted before mining and/or processing could be established within a regulatory environment that is appropriately protective of the health and safety of workers, the public, and the environment. There is only limited experience with modern underground and open pit uranium mining and processing practices in the wider United States, and no such experience in Virginia.[6]

The arguments against mining Coles's deposit were largely geophysical: primarily climatic, with an additional tectonic component. The body of the text refers 26 times to "extreme" climate-related events, and this notion of "extremes" captured the public discourse as a reason to keep the moratorium in place. In town hall meetings, newspaper editorials, and news stories, the specter of contamination of the water supply of Virginia Beach (175 air miles distant from the nearest economically viable uranium deposit) by a flood-induced impoundment failure of modestly radioactive mine tailings was raised repeatedly. Uranium tailings are the leftovers after mining, and they need to be held in place for centuries.

In *UMV*, the National Research Council made little attempt to compare its geophysical observations to conditions in other areas of the country, or to the parts of the world where uranium mining or processing has taken place or is being done successfully. Further, the report makes no attempt to differentiate the localized Coles Hill climate and seismic risk from those of a general survey of the state. While the National Research Council largely confined its climatic analyses to Virginia, virtually every location in the United States can claim some history of extreme rainfall, wind, snowfall, or drought. In reality, the climate extremes expected at Coles Hill are much less than those experienced elsewhere where uranium is mined.[7]

Chapter 2 of *UMV* is "Virginia Physical and Social Context," and it contains many references to floods and tropical cyclones. According to the report:

Virginia is subject to extreme weather events—hurricanes and tropical storms, thunderstorms, and heavy rainfall and snowfall. In the period from 1933 to 1996, 27 hurricanes and/or tropical storms made landfall in Virginia, bringing with them the threats of flooding, high winds, and tornadoes.[8]

The relevant citation is an obscure publication from the National Weather Service Wakefield, Virginia, office, titled "Historical Hurricane Tracks, 1933–1998, Virginia *and the Carolinas*" [emphasis added],[9] yet a direct examination of the historical tropical cyclone tracks from the National Hurricane Center reveals that only eight tropical storms (and no hurricanes) made landfall in Virginia (as opposed to the "27 hurricanes or tropical storms" *UMV* claimed) from 1933 through 1998. Nineteen were in North and South Carolina.[10] *UMV* is clearly misleading in its documentation of Virginia's tropical cyclone history, an inaccuracy that leads to the assessment of a greater risk than is actually present in the state's existing climate.

It was also inappropriate to consider the entire state as a whole. Instead, Figure 8.1 shows tropical cyclone tracks within 160 kilometers (100 miles) of Coles Hill for 1930 through 2011. There was one category 3 hurricane (1954's Hurricane Hazel) and two category 1 hurricanes; all of the other passages were either tropical storms or tropical depressions. The intensity of tropical cyclone passages within 160 kilometers of the Coles Hill site is considerably lower than that of tropical cyclones making landfall along the Virginia coast.

In the subsequent history of extreme Virginia floods, *UMV* repeatedly refers to 1969's Hurricane Camille, which, though not a hurricane in Virginia, produced one of the most intense rainfalls ever measured in the coterminous United States. Camille came ashore around midnight on August 16–17 near Pass Christian, on the western Gulf Coast of Mississippi as only the second category 5 hurricane of the 20th century. Camille moved north across the state and maintained hurricane strength for an unusually long time. After 14 hours it weakened to a tropical storm and, as it tracked east across Kentucky, it was a tropical depression.

By then, Camille was producing unremarkable rain totals of one to three inches as it moved eastward. But that all changed on the evening of August 19, when it collided with some of the highest-relief mountains in the Appalachians. Camille's moisture was forced skyward as a cold front from the northwest intersected it over the mountains, providing additional lifting.

According to *UMV*, it "produced heavy rainfall of up to 790 mm (31.1 in) as it crossed the state." The definitive study of the Camille rainfall states:

> The greatest amount, 27 in., was measured near Massies Mill, Va. The Weather Bureau received a report of 31 in. of rain in 5 hr measured at the junction of the Tye and Piney Rivers. Since the timing of rain did not agree with other nearby reports and the catch, if verified, might be a world record, the site was visited by a Weather Bureau representative to more fully document the event. Neither the person who made the observation nor the container that was used could be identified. Therefore, the 31 in. was not

Tropical cyclone tracks within 100 miles of Kingsville Dome (top) and Coles Hill (bottom), 1930–2011

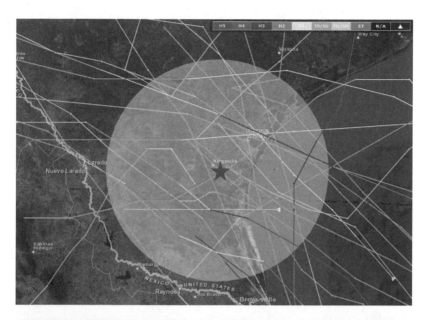

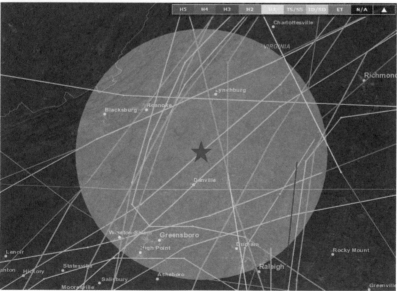

Source: National Oceanic and Atmospheric Administration, "Historical Hurricane Tracks," 2013, http://www.csc.noaa.gov/hurricanes/#.

Note: There were 27 tracks around Kingsville Dome and 26 at Coles Hill, but the storms were much stronger in Texas.

used. However, a reliable measurement of 23 in. was obtained for the same general vicinity, indicating that the rainfall was extreme, and that 31 in. may have fallen.[11]

Subsequently, it was calculated that the return interval of a flood of the magnitude initiated by Camille in the Blue Ridge and Appalachian Mountains in Virginia is 2,000 to 4,000 years.[12]

UMV also focuses on what is known colloquially as the 1995 Madison County flood, a terrain-focused regenerating thunderstorm complex that produced 600 millimeters (23.6 inches) of rain in six hours, and was accompanied by "more than 500 separate landslides, debris flows, and debris avalanches."

Conflating these two events with possible flooding and slope failure at Coles Hill ignores the fact that both of these events were topographically dependent, that the rolling, low-elevation terrain around Coles Hill is simply incapable of significantly boosting rainfall, and that the local terrain does not have high enough relief to produce substantial debris flows. This is obvious from Figure 8.2, which depicts the relative topography of both of these flood sites, along with the region around Coles Hill.

UMV then notes that 1996 Hurricane Fran (which was Tropical Storm Fran in Virginia) brought 400 millimeters (15.6 inches) of rain, without mentioning that the nearest weather station to Coles Hill, in Chatham, Virginia, received only 124 millimeters (4.88 inches). Hurricane Irene in 2011, also noted in UMV, never made landfall in Virginia and was of no consequence in the Coles Hill region. It would seem impossible to claim these events were significant to Coles Hill after checking the (readily available) observed data.

A more accurate and specific analysis indicates that an extreme flood-producing rainfall event is considerably less likely for the Coles Hill location than for other specific locations across the Commonwealth of Virginia. Tropical cyclones are much stronger closer to the Atlantic coast, and terrain-induced floods are more severe farther to the west (see Figure 8.1).

UMV's perseveration on tropical cyclones in Virginia means that their effects must be among the "steep hurdles to be surmounted

FIGURE 8.2
Google Maps topography showing that the relief for the region of the Madison County flood (left) and Hurricane Camille (center) was dramatically different from the relief around Coles Hill (right)

Graves Mill, Madison Co., Va.   Massies Mill, Nelson Co., Va.   Coles Hill, Pittsylvania Co., Va.

prior" to the exploitation of the Coles Hill site. What *UMV* conspicuously fails to mention is that uranium has been successfully mined and processed in other places with much more severe tropical cyclone climatologies. These include south Texas, where the Kingsville Dome uranium operation is located only 40 kilometers (24.9 miles) from the Gulf of Mexico, two sites in the former French colony of Madagascar (Tranomaro and Folakara) from which uranium was exported in large quantities, and the Ranger Uranium Mine site in Australia's Northern Territory, one of the largest uranium mines (in volume) on earth.

More frequent and stronger tropical cyclones during the period from 1930 through 2011 passed within 100 miles of the active and large Ranger mine in Australia (63 total storms) and over the previously mined region around Folakara, Madagascar (68 total storms).[13] Ranger is an open-pit design that is shut down when tropical cyclones threaten or deliver very heavy rainfall.

As another "steep hurdle," *UMV* provides a lengthy list of Virginia floods (none of which affected Coles Hill), but it neglects to place them in a climatological perspective. Figure 8.3 shows the probable maximum precipitation (PMP) distributions around the coterminous U.S. east of longitude 105°W for 24-hour rainfall over 100 square miles. Again, the climatic environment is much more extreme in south Texas, where the PMP is near the highest in the coterminous 48 states.

FIGURE 8.3
Probable maximum precipitation for 24 hours at the 100-square-mile level

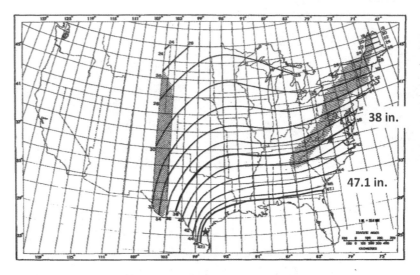

Source: National Oceanic and Atmospheric Administration, "Probable Maximum Precipitation Estimates, United States East of the 105th Meridian," Hydrometeorological Report HMR-51 (Washington: Department of Commerce, 1978).

Note: It is obvious that the flood potential at the Texas site is much greater than that at Coles Hill (the two blue stars are the Kingsville and Coles Hill sites).

Given the probable size and design of a Virginia mine, one can calculate how much rain would have to fall to generate a flood that would affect the tailings. That amount is approximately *twice* the PMP at Coles Hill today, or approximately 76 inches. For perspective, the 24-hour rainfall required to flood a prospective tailings deposit is slightly more than the *world record* 24-hour rainfall of 71.8 inches measured at Reunion Island on January 8, 1966. Reunion is approximately 800 miles east of the Folakara uranium mining district in Madagascar, and mining operations began around the time of the record rainfall. Yet Folakara is not mentioned once in *UMV*.[14]

The geophysical sleight of hand in *UMV* was not limited to selective and misleading climatological analyses. After the alarming rhetoric about rainfall and flooding, it switched to geology—specifically earthquakes—emphasizing potential seismic hazards in Virginia.

There are two active seismic zones in Virginia where small to moderate earthquakes occur with some frequency. As shown in Figure 8.4, Coles Hill is in neither. Within the Central Virginia Seismic Zone (CVSZ), a 5.8 magnitude earthquake occurred in August 2011. Shaking as a result of this earthquake was felt across the eastern United States, most strongly near the earthquake's epicenter near Mineral, Virginia. This was the strongest recorded earthquake in the CVSZ, exceeding by 10 times the strength of the previously recorded strongest quake—a 4.8 magnitude earthquake there in 1875. The strongest earthquake ever recorded in Virginia was a 5.9 magnitude earthquake in May 1897 in the Giles County Seismic Zone located in southwestern Virginia, hundreds of miles away from Coles Hill.

Earthquake maps prepared by the United States Geologic Survey (USGS) show that the earthquake risk in Virginia is relatively low in comparison with that in other areas of the United States; and at a more local level, the potential activity at the Coles Hill location is lower than in other portions of the state. Based on models from the USGS-National Seismic Hazard Mapping Project, the probability of an earthquake of the magnitude that occurred in the CVSZ in August 2011 occurring within 50 kilometers (31 miles) of Coles Hill is less than 0.5 percent per century (a less than 1 in 20 chance per 1,000 years).[15]

Coles Hill is geophysically unremarkable in comparison with other regions where uranium has been successfully mined or processed. Moreover, *UMV* should have noted that the seismic "natural hazard" in Virginia is insignificant in the region surrounding Coles Hill.

FIGURE 8.4

Virginia regional seismic potential

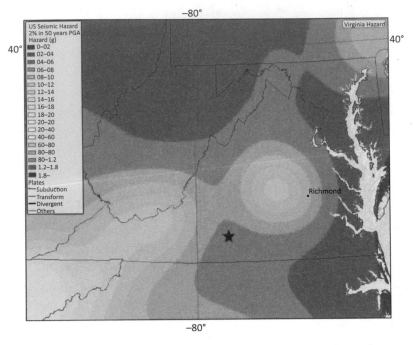

Source: United States Geological Survey, "USGS National Seismic Hazard Maps, Data, and Documentation," 2008, http://earthquake.usgs.gov/hazards/products/graphic2pct50.jpg.

Note: Medium to deep blue indicates a very low potential.

Conclusion

The root cause of the biases in *UMV* lies in the power of the government-created National Academy of Sciences to continue to do what it was chartered to by President Lincoln: provide support for government policies, be they winning a civil war or a war on uranium-powered nuclear energy. The panel that produced *UMV* knew full well the level of commitment of the Obama administration to so-called renewable power sources. New nuclear capacity, while not as cost-efficient as natural gas for electrical generation, is dramatically more dependable and rich in power than the administration's choices, and would consign renewables to marginal power production with geographically limited application.

The rhetorical and logical excesses of *UMV* are remarkably easy to spot, and were just as easy for anti-nuclear activists to exploit. It is doubtless that the *UMV* committee knew precisely what would happen, and that their misleading geophysical arguments would result in the Virginia legislature turning down any attempt to rescind the existing moratorium on uranium mining in the state.

While activists gained, the losers were the Coles family, the stockholders of Virginia Energy Resources (the latest iteration of Virginia Uranium), the economically depressed area surrounding Coles Hill (Pittsylvania County per capita income in 2010 was $20,468), and the credibility of National Research Council. It selected a committee that, based on professional history, would likely make it virtually impossible to surmount the *political* hurdles standing in the way of the exploitation of the largest U.S. deposit of a very clean and efficient fuel. This was frank and politically correct confiscation, based on a systematically misleading analysis of irrelevant geophysics. The Virginia Uranium fiasco shows what can happen when the weight of science becomes the abuse of authority. In this case, $7 billion was confiscated from the Coles family.

Epilogue

On May 21, 2018, the Supreme Court granted a writ of *certiorari* in the case *Virginia Uranium v. Warren*, in which Virginia Uranium argued that the Virginia moratorium on uranium mining preempts what is rightfully the purview of the federal Atomic Energy Act. The federal district court held that the Atomic Energy Act applies only to uranium on public lands, and the Fourth Circuit court of appeals upheld the district court's dismissal of Virginia Uranium's petition. The Supreme Court heard public arguments on the case on November 5, 2018. On June 17, 2019, the Supreme Court ruled 6-3 against Virginia Uranium, with the majority holding that there was no specific pre-emption of state uranium mining regulation (including a moratorium). In a subsequent public comment, Virginia Uranium said:

We continue to think that Virginia's uranium mining ban is both unlawful and unwise, and we are reviewing other options for challenging the Commonwealth's confiscation of Virginia Uranium's mineral estate.

It would appear that Virginia Uranium's next move will be to claim a massive regulatory taking of several billion dollars. If they valued it at its original $7 billion, depending upon the settlement, this could turn out to be the largest civil taking in U.S. history.

CAN POLITICS TURN GOLD INTO DROSS?

The Story of Alaska's Pebble Mine

ere is another example of the abuse of power in the name of "science," with an important twist. A federal regulatory agency, the Environmental Protection Agency (EPA), works hand-in-hand with those opposed to development of what is currently thought to be the largest known copper, gold, and molybdenum deposit on earth. It is on state land zoned by the state of Alaska for mining. The deposit is beneath land that is drained by a small creek that empties into pristine Lake Iliamna. That lake ultimately drains into Bristol Bay—almost 100 miles away—and home to the world's largest wild sockeye salmon fishery. The Pebble story brings politics, environmental policy, and economic development into close proximity, and is (to date) a cautionary tale on how a powerful regulatory agency can go rogue.

As the chapters in this book have described, there are several ways in which science can be corrupted. The incentive structure certainly results in a canon of knowledge that is massively littered with false-positive results. But another problem might be termed "corruption by authority." In this book we describe how the National Academy of Sciences, through its National Research Council, can assemble apparently impartial and unbiased expert panels to study a subject, but in fact the results of the study are largely known beforehand because of the track records of the chosen participants. In this manner, it's easy to suppress dissenting views even when their proponents are on the panel. That is clearly what happened with Walter Coles Jr. and Virginia Uranium, Inc. (as described in Chapter 8).

With Pebble, the EPA colluded with environmental organizations to create what could only be termed a science fiction to prevent the exploitation of a world-class copper and gold deposit. The deposit was discovered in 1987 by Cominco, a Canadian mining company headquartered in Vancouver, British Columbia. The "Pebble project" is located in southwestern Alaska, on state lands that were accepted by Alaska as part of a land swap with the federal government, specifically with the land's mineral and mining potential in mind. In addition, the site has been designated through two public land-use planning processes for mineral exploration and development.[1]

But this story is not about what *will* happen at the Pebble mine site, if it is ever approved. This is about regulatory agency policies and actions that have affected the Pebble proposal. It is a remarkable tale of agency overreach, unethical behavior, intrigue, and what can only be called "fake science." The Pebble story is yet another example of empowered agencies and entities bending (and sometimes creating) science with the sole purpose of executing a policy. In this case, as with Virginia Uranium, it also entailed a mine.

With Pebble, the EPA, environmental nongovernmental organizations (NGOs), and Alaska's powerful environmental lobby managed to circumvent landmark legislation, the 1970 National Environmental Policy Act (NEPA), that empowered President Nixon to create the EPA itself. NEPA is the government's main decision "tool" for determining the disposition of all proposed projects that could have a significant environmental impact on air, land, water, and human health and welfare. The relevant question for Pebble is: Why did the EPA circumvent NEPA, whose process is quite clear and explicit, rooted in the law itself, from which equal justice is supposed to emanate?

The Council on Environmental Quality (CEQ) or the EPA determines what activities require a formal environmental impact statement (EIS). Clearly, a substantial mine in sparsely populated country near federal wilderness areas, such as Pebble, would (and should) be required to be reviewed under a formal EIS. Small streams that drain the Pebble property eventually flow into Lake Iliamna, by volume the seventh largest U.S. lake. The lake ultimately drains into Bristol Bay, home to the world's largest sockeye salmon fishery. But the creeks that flow through the Pebble site contribute a very small increment of water in comparison with all the other drainages into the massive bay, which is approximately 100 miles away.

Instead of having Pebble proceed with its own EIS, the EPA substituted its own assessment of the impact of the Pebble project on the Bristol Bay Watershed (BBWA) for the formal and comprehensive NEPA EIS decision process.

The BBWA appeared in the form of a report. In May 2012, the EPA issued a review draft titled "An Assessment of the Potential Mining Impacts on Salmon Ecosystems of Bristol Bay, Alaska." On April 20, a share of Northern Dynasty Minerals (NYSE: NAK) traded at $5.80, and the stock was considered a fairly conservative investment, and certainly a staple in many Canadian retirement accounts. After discovering the massive deposit, NAK acquired major financial backing from one of the world's largest mining concerns, Anglo-American, which invested over $500 million into startup expenses for NAK. By May 25, a NAK share sold for $2.48. The *draft* EPA report had stripped nearly 60 percent of the stock's value in a month.

According to the EPA, it became involved in the permitting of the project because of petitions against the mine from Native Alaskan tribes in 2010. But oral statements from EPA employees and official agency documents reveal the existence of an internal EPA "option paper" that make clear the agency opposed the mine on ideological grounds and had *already decided* to veto the proposal in the spring of 2010. The draft Bristol Bay report was not released until two years later.

Much of this information was found through legal discovery in a case that Pebble brought against the EPA, alleging that the agency had violated the Federal Advisory Committee Act (FACA) by colluding with anti-Pebble activists to preemptively prevent Pebble from even applying for a permit to mine, which is an integral part of the NEPA process. That discovery indicates beyond doubt that the Pebble project was being denied access to the well-established and accepted NEPA process.

The Importance of Pebble

The Pebble project has the potential to supply as much as one-quarter of the United States' copper needs over more than a century of production, in addition to large quantities of gold, silver, molybdenum, and other minerals.[2]

There are several important stakeholders; environmentalists, sportsmen, and the fishing industry are concerned that mining the deposit will despoil Bristol Bay, home of the world's largest sockeye salmon fishery. Investors in the Pebble Partnership obviously want the site developed, which has been zoned for mineral development by the state of Alaska. In addition, there are local economic issues to consider. Substantial unemployment in south coastal Alaska is endemic, and, according to Pebble, the original and related construction and support activity would provide around 15,000 jobs and contribute more than $2.5 billion to the country's GDP each year. A 2017 revised proposal is for a somewhat smaller enterprise, with the jobs and GDP numbers varying accordingly.

Opposition to the Pebble mine is part of a larger "leave it in the ground" movement, and Pebble has become a proxy for undesired mining projects. For example, in 2014, the EPA was under pressure from a Native American tribe to veto an iron ore mine in Iron County, Wisconsin. Similarly, an environmental group in Minnesota lobbied against a nickel-platinum-palladium mine in the northeastern part of the state. The EPA is also being urged to veto a planned nickel mine in Oregon near a tributary of the Smith River.[3]

However, the details of these four mining projects, including Pebble, are still on the drawing board, and they have not gone through the normal NEPA environmental impact analysis. Writing in the *Wall Street Journal*, mineral resource expert Daniel McGroarty noted:

> What the Wisconsin, Minnesota, and Oregon mine projects have in common is that none has put forward an actual mine plan.

Neither has Pebble. Submitting a mine plan would trigger a thorough mine plan review as required under NEPA (the National Environmental Policy Act enacted by Congress in 1970). For more than 40 years NEPA has defined the process by which a mine or any other resource project is evaluated. Under the law, every one of the concerns raised by the opponents to the Wisconsin, Minnesota, and Oregon mines would be aired as public comments, and examined by scientists and technical experts, before approval is granted or denied. Using the Pebble mine as precedent, anti-mining activists are urging the EPA to ignore NEPA and bar mining projects with no review necessary.

. . .

Current law requires an environmental impact statement which is an extensive assessment of the mine's potential impact weighed against mitigating safeguards. But anti-mining activists are pushing for a switch to "cumulative effects assessments," which would take into account past, present and future actions in the project vicinity. Under such an approach, a mine could be vetoed because other proposed mines in the region could *at some point* in the future collectively contribute to deleterious environmental effects. Even the most meticulously engineered mine plan can be undone by a parade of hypothetical horribles.[4]

Indeed, the EPA designed a fictional Pebble mine in its 2014 *Bristol Bay Watershed Assessment*, which it then used to preempt the real Pebble proposal under the Clean Water Act.

Clean Water Act Invoked to Halt Pebble

The Clean Water Act (CWA) was passed by Congress in 1972. It establishes the basic structure for regulating pollutant discharges into the waters of the United States, giving the EPA the authority to implement pollution control programs such as setting wastewater standards for industry. In 2010, before Pebble even submitted an application for a mining permit, the EPA used a specific provision of the Clean Water Act known as Section 404(c), to preempt the mine permit application. According to the act, the Pebble Partnership is entitled to apply for a permit and the U.S. Army Corps of Engineers has the responsibility to approve or disapprove the application. However, in a clearly unintended consequence, the EPA veto called into question the legality of preempting the issuance of a permit before the permit application had been submitted for review, as required under the act, because it was based on a fictional, worst-case mine design that originated within the agency itself.[5]

How was Section 404(c) used to halt the Pebble project? According to the EPA, the act authorizes the Army Corps of Engineers through Section 404(a), or an approved state through Section 404(h), to issue permits for discharges of dredged or fill material at specified sites in waters of the United States. Section 404(c), however, authorizes the EPA to restrict, prohibit, deny, or withdraw the use of an area as a disposal site for dredged or fill material if the discharge will have unacceptable adverse effects on municipal water supplies, shellfish beds and fishery areas, wildlife, or recreational areas. The EPA believes it has "veto authority" under Section 404(c), and may initiate a public process to prohibit or restrict the specification by the Corps of Engineers or by a state, for the discharge of dredged or fill material at a particular site.[6]

According to the Clean Water Act, Section 404(c) authority may be exercised before a permit is applied for, while an application is pending, or after a permit has been issued. Because most Section 404(c) actions have been taken in response to unresolved Corps of Engineers permit applications, this type of action is frequently referred to as "an EPA veto of a Corps permit." Although the Corps of Engineers authorizes approximately 68,000 permit activities in U.S. waters each year, the EPA has used its Section 404(c) authority very sparingly, exercising it thirteen times in the forty-plus-year history of the act, with only two determinations being made in the past twenty years. There are eleven instances in which Section 404(c) denials were issued from 1980 to 1991, then none for almost two decades, until Pebble.[7]

Although used sparingly, the EPA's authority under Section 404(c) is well tested in the courts. District courts have overturned (reversed) such determinations on a variety of project-specific grounds; however, those reversals of the EPA's determinations did not survive the appeal process. Legal opinions vary, but most agree that "avoiding a withdrawal of the waters at issue under 404(c) may be the best plan that the Pebble Partnership has in keeping its project alive. It has been easier for Pebble to defend such a decision in court rather than challenge an adverse decision made by EPA."[8]

Bristol Bay Assessment Crafted to Kill Pebble

The EPA claims that its 2014 veto of Pebble under Section 404(c) was based on "scientific evidence" presented in the BBWA, commissioned by the EPA in February 2011. After producing two draft versions (in 2012 and 2013), the final BBWA was published in 2014, supposedly to present the science behind the potential impacts of Pebble on Bristol Bay.[9]

However, because there was never any mining permit application, and therefore no submission of a mine plan design, the EPA charged a senior biological scientist named Philip North to design a worst-case scenario, an open-pit "hypothetical mine" that would have no chance of being approved in a review by a professional mining engineer. In fact, Pebble's real intentions for mining the deposit and its mine plan design have never been completely disclosed, although in late 2017 Northern Dynasty announced that only a portion of the deposit would be exploited.[10] Nevertheless, the EPA continued to rely on North's hypothetical mine design and its fictional impacts.[11]

The Pebble Partnership knew it would be required to file a detailed application for a permit, and that an EIS for the entire proposed mining operation would need to be produced by the lead federal agency. Consequently, it spent approximately $150 million and nearly ten years compiling a massive database of the geology, biology, ecology, and dynamics of the Bristol Bay watershed. The EPA and North simply ignored this comprehensive repository of information.[12] Both North and other EPA officials have admitted under oath that during the entire time that the BBWA was being written (2011–2014), they never really intended to provide a scientific foundation for regulatory decisionmaking.[13] One is tempted to ask the question: Then why did the EPA design a fictional mine?

There is more to the story. While creating a fictional open-pit mine, the EPA's North also coached anti-Pebble activists on how to petition the agency to stop any mine permit application. In fact, it appears he wrote the petitions. When these actions surfaced in early 2013, the U.S. House of Representatives Oversight Committee asked to speak with North about his role at the EPA in the Pebble application. His response was to flee the country, after which a federal judge issued a subpoena in August 2015 directing North to appear before the House committee. He was finally served subpoena papers in Australia in January 2016 and was deposed in April 2016 by attorneys for the Pebble Partnership and staff attorneys from the House Committee on Science, Space, and Technology.[14]

There are nearly ten years of emails and internal memos that indicate collusion between EPA officials and environmental activists opposed to Pebble—much of which was produced from Freedom of Information Act (FOIA) requests from Northern Dynasty. The EPA's Region 10 administrator for Alaska, Dennis McLerren, was deposed by the House committee in 2016 because he was thought to have played some role in Pebble's application denial. Much has been learned through discovery pertaining to subsequent Northern Dynasty FOIA inquiries to the EPA about how individuals within the agency handled the Pebble project.[15]

The weight of evidence mounting from depositions and FOIA requests about the absence of impartiality in the EPA's adjudication of the Pebble proposal over many years finally reached a critical stage. In 2015, attorneys representing Northern Dynasty petitioned the EPA Office of the Inspector General (OIG) to conduct an investigation concerning the BBWA. Northern Dynasty's petition made several powerful evidence-based points that had been uncovered by FOIA requests and in House committee proceedings:

1. The EPA's seeking to veto Pebble did not originate from complaints of federally recognized tribes in Alaska, but came from within the agency itself:

   This evidence, obtained from the EPA under FOIA, suggests that EPA officials in Alaska began musing about the potential for a preemptive 404(c) veto of the project, and lining up other federal agencies to support this plan, some two years before the first petition was received from federally recognized tribes.

2. The EPA's BBWA was designed to support a veto, rather than being an objective inquiry:

The Assessment evaluated a mine scenario coauthored by Mr. North (the EPA's principal early advocate for a veto of the Pebble project) who publicly admitted that he did not include state-of-the-art technology because he assumed that mining companies would not use what is available. This critical flaw was recognized by numerous independent peer reviewers (selected by the EPA), who said precisely the opposite—that the permitting process would require much more and better technology than what the EPA used for its Assessment. The BBWA uses a mine scenario that fails to meet *legal requirements* to protect against harm to salmon, by assuming a fictional mine that does not meet modern standards for environmental protection.

3. The EPA biased the peer-review process:

The EPA manipulated the peer review of the BBWA itself in a way designed to minimize criticism. The EPA violated its own standards when, during the first peer review, it unduly restricted the schedule, shielded the peer reviewers from public comments, and then held a closed-door meeting with the peer review panel. During the second peer review, the EPA shut out the public entirely, completely violating its own standards for transparency.[16]

In summary, Section 404(c) of the Clean Water Act was used to halt the Pebble project from moving forward, but the BBWA was used to attempt to kill the project outright because, according to the EPA, it is based on "science." The mine plan fabrication is an egregious example of federal agency deception and distortion of science reported in a "scientific assessment." The application of this process to deny a person or a corporation of its property rights is hardly unique, as shown in the Virginia Uranium story (Chapter 8).

The House Oversight Committee, in a November 4, 2015, letter to the EPA administrator, characterized the agency's actions regarding Pebble's rights under NEPA as "highly questionable and lacking legal basis," and urged the administrator to "allow the project proposals to go forward under the Clean Water Act and National Environmental Policy Act (NEPA)." The EPA's preemptive veto of the Pebble project has a deeper meaning that should disturb environmentalists much more than the proposed mine: the veto preempted the NEPA process—the Magna Carta of environmental laws—from being triggered to study the mining proposal in detail, as thousands of proposals have been studied over the past 45 years. The EPA appears to have issued its veto to avoid the "risk" of a possible NEPA-approved mining operation that it did not want, and the discovery process has clearly borne that out. The EPA set a very negative precedent by circumventing NEPA, which is responsible for its very existence.[17]

Officially, NEPA applies whenever a proposed activity or action

- is proposed on federal lands,
- requires passage across federal lands,
- will be funded in part or in whole by federal money, or
- will affect the air or water quality that is regulated by federal law.[18]

Concerns about the EPA side-stepping the NEPA process for the Pebble proposal were expressed by other stakeholder federal agencies, such as the U.S. Geological Survey (USGS). One of the USGS's scientist contributors to the BBWA stated the following:

The thing that has always bothered me about the assessment [BBWA] is that there is a mechanism in place to review mine permit applications [the NEPA process]. The process was created by EPA, yet the decision was made by EPA to short-circuit their own process and explore a 404(c) veto action. . . .

From my perspective, Northern Dynasty and the Pebble Limited Partnership acted in good faith and went well beyond what would be considered standard practice for a mine permitting exercise anywhere in the United States or in the world. I took their extraordinary effort to reflect their appreciation of the sensitivity of the environment where they are working.

. . .

The NEPA process seemed to be working perfectly fine at Pebble and I see no reason why the NEPA process should not be allowed to render a final verdict rather than having this other path bar it.[19]

Because of the irregularities noted above, a district court judge in May 2014 issued a preliminary injunction against any further efforts by the EPA to deny Pebble its due process rights to develop and submit a permit application.

Based on congressional inquiries and political pressure, the EPA decided to conduct an internal review regarding its conduct during the BBWA process. The EPA leadership charged the agency with determining whether it had conducted the BBWA in a biased manner and predetermined the outcome, or followed policies and proper procedures for ecological risk assessment, peer review, and information quality. Based on available information, the EPA Office of Inspector General claimed to have found no evidence of bias in how the agency conducted its assessment, or that the BBWA team members, or agency leadership, predetermined the assessment outcome.

On January 13, 2016, the EPA published the OIG's findings on how it conducted the assessment in the three primary phases discussed in the agency's ecological risk assessment guidelines. The review indicated that the EPA's work on the assessment met requirements for peer review, provided for public involvement throughout the peer review process, and followed procedures for reviewing and verifying the quality of information in the assessment before releasing it to the public.[20]

The EPA Stands by Its Bristol Bay Study

The EPA's OIG review was prompted by a request from the Pebble Partnership, the state of Alaska, and other parties to investigate allegations of bias, predetermination of outcomes, inappropriate collusion with special interest groups and other process abuses with respect to the EPA's BBWA, and subsequent regulatory action to preemptively veto the Pebble project under Section 404(c) of the Clean Water Act. While acknowledging significant "scope limitations" in its (EPA) review and subsequent report, the OIG concluded: "We found no evidence of bias in how the EPA conducted its assessment of the Bristol Bay watershed, or that the EPA pre-determined the assessment outcome," but that an EPA Region 10 employee may have been guilty of "a possible misuse of position."[21] This, of course, was the infamous Philip North.

Several previous investigations of EPA conduct toward Pebble contradict the OIG report. The House Committee on Oversight and Government Reform found "that EPA employees had inappropriate contact with outside groups and failed to conduct an impartial, fact-based review of the proposed Pebble mine." Former U.S. senator and secretary of defense William Cohen, who produced his own analysis as a consultant to Northern Dynasty, said his investigation "raise[s] serious concerns as to whether EPA orchestrated the process to reach a pre-determined outcome; had inappropriately close relationships with anti-mine advocates, and was candid about its decision-making process."[22]

After the EPA published its internal review of the BBWA process, the Pebble Partnership countered with a response to the EPA's OIG report in January 2016. It is the Pebble Partnership's view that the OIG investigation into EPA misconduct was so narrow as to materially distort the reality of the agency's actions. Further, it is Pebble's view that the "possible misuse of position" cited by the OIG with respect to an EPA employee in Alaska underestimates the seriousness of agency misconduct and diminishes accountability for the misconduct to a single individual despite evidence that senior EPA staff at Region 10 (Seattle) and at headquarters in Washington, DC, were aware of and complicit in inappropriate activities.[23]

A cursory review of the scope of the OIG investigation demonstrates why it was unable to expose EPA misconduct with respect to the BBWA and subsequent efforts to veto the Pebble project. Despite more than 100 EPA employees playing a role in the agency's efforts to preemptively veto Pebble, the OIG reviewed the emails of only three EPA officials.

And despite the close collaboration of dozens of anti-mine activists in the EPA's actions regarding Pebble, the OIG reviewed the emails of only one anti-mine activist.

The EPA's BBWA study process was initiated in February 2011 and concluded in January 2014, and the agency's Section 404(c) veto was initiated in February 2014 and suspended in November 2014 following a preliminary injunction issued by a federal court judge; however, the OIG reviewed EPA emails only through May 2012. During the two and a half years of activity unexamined by the OIG, the EPA issued two more versions of the BBWA, including its final report, conducted multiple disputed peer review processes, and initiated its preemptive 404(c) veto.[24]

Philip North was found to have no emails available for a 25-month period of time within the OIG's already limited 52-month window of investigation. The OIG did not seek to recover any emails from three key EPA officials. Indeed, their emails may have been deleted before the onset of its investigation. Rather than review all retrieved emails, the OIG used undisclosed search terms to further narrow its review. Finally, the OIG did not seek records from the private email accounts of EPA officials, despite evidence that officials used private email accounts to conduct government business, against all federal employee protocols.[25]

Despite its wide-ranging investigative authority, the OIG issued just one subpoena with respect to its Pebble review. That subpoena, issued in August 2015 by a federal judge to counsel for a former EPA official (Philip North), was summarily ignored.[26] To force investigation of EPA actions by the OIG, Pebble Partnership reviewed more than 50,000 documents received via FOIA requests and submitted a total of 19 letters spanning 214 pages and appending nearly 600 exhibits.

The FOIA requests to the EPA addressed a wide range of concerns about the agency's actions, presented corresponding evidence, and called on the OIG to use its subpoena powers and other authority to more fully investigate EPA actions involving Pebble. While the OIG report finds no evidence of bias or predetermination of outcomes with respect to the BBWA, it provides no findings at all on a large number of other important matters, such as the collusion between North and the Native tribes with regard to a preemptive veto. Nor does the OIG report comment on the evidence provided by Pebble Partnership in raising its concerns.

When the OIG charged North with the vague "possible misuse of position," there was no mention of his use of a private email account in 2011 to coordinate with an anti-mine activist in the preparation of a tribal petition that the EPA cited as the sole catalyst for its BBWA study and preemptive 404(c) regulatory action.[27]

The OIG report found that North acted alone in this collusion, and that the "employee's supervisor told us that he was not aware that the employee had taken such an action." However, the OIG fails to note the many other substantive interactions North had with anti-mine activists or the extent to which this collusion was known throughout the agency. Evidence uncovered by the Pebble partnership shows that at least six EPA employees knew about the improper collusion between North and anti-mine activists. As early as 2010, at least two EPA employees alerted senior EPA staff and an EPA attorney about these inappropriate contacts, but no corrective action was taken.[28]

Congress has authority to provide oversight for inspectors general where an inspector general fails to uncover or report clear misconduct on the part of an agency, and it should be doing so now.

Final Chapter of the Pebble Story—For Now

This history of Pebble begins in 2005, before both the Trump and Obama administrations. In 2016, market perception was that the incoming Trump administration intended to reverse some of the previous administration's opposition to the mine. Northern Dynasty stock doubled in price between the November 2016 election and the turn of the year.

But it was Congress that intervened first. On February 22, 2017, Rep. Lamar Smith (R-TX), chair of the Science, Space and Technology Committee, wrote to the EPA administrator, Scott Pruitt, urging the agency to normalize the permitting process for Pebble:

> The Committee recommends that the incoming administration rescind the EPA's proposed determination to use Section 404(c) in

a preemptive fashion for the Pebble mine in Bristol Bay, Alaska. This simple action will allow a return to the long-established Clean Water Act permitting process—along with NEPA—and stop attempts by the EPA to improperly expand its authority. Moreover, it will create regulatory certainty for future development projects that will create jobs and contribute to the American economy.[29]

President Trump signed an Executive Order on February 28, 2017, directing the Environmental Protection Agency to revise its expansive interpretation of the "waters of the United States" definition, in which ditches and drainages would often be treated as "navigable."[30] On March 28, 2017, he signed an Executive Order lifting a ban on new coal mining leases on federal land.

Then, on May 11, 2017, the EPA entered into an agreement with Pebble Limited Partnership that ended ongoing litigation and put Pebble squarely on the path to the standard NEPA permitting process described below. According to the EPA, the out-of-court settlement with Pebble would be "to resolve litigation from 2014 relating to EPA's prior work in the Bristol Bay watershed in Alaska." The EPA stated that the "settlement provides Pebble an opportunity to apply for a Clean Water Act (CWA) permit from the U.S. Army Corps of Engineers before EPA may move forward with its CWA process." Finally, the agency affirmed, "The agreement will not guarantee or prejudge a particular outcome, but will provide Pebble a fair process for their permit application and help steer EPA away from costly and time-consuming litigation."[31]

On January 26, 2018, Pruitt issued a surprise statement concerning Pebble:

It is my judgment at this time that any mining projects in the region likely pose a risk to the abundant natural resources that exist there.

In an accompanying press release, the EPA said:

This decision neither deters nor derails the application process of Pebble Limited Partnership's proposed project. The project proponents continue to enjoy the protection of due process and the right to proceed. However, their permit application must clear a high bar, because EPA believes the risk to Bristol Bay may be unacceptable.

One can only take the EPA at face value here. It allows Pebble to complete its application with the Corps of Engineers, but it appears to be saying that it retains the right to veto any application under 404(c) after all. While all of this may appear clear as mud to us mortals, the investing public saw a grim portent. By March 8, 2018, NAK had lost 40 percent of its January 26 valuation.

Looking back over the saga of Pebble, it is clear that the EPA had behaved badly during the previous administration. And, given that the Pebble mine remains a political hot potato, what it will do in the future is not at all clear. What is clear, though, is that rumors that the Trump administration would rubber stamp the Pebble mine permit were premature.

Even with its future path uncertain, however, the EPA's creation of a fictional mine to drive the BBWA report is testimony to the ability of the federal government to manipulate science for political ends.

Postscript: On November 6, 2018, Alaska voters in a statewide election elected a new governor, Republican Mike Dunleavy. According to Reuters on November 7, Governor-elect Dunleavy, a former state senator, is a strong proponent of encouraging investment and responsible development of the state's natural resources to address the significant fiscal and economic challenges facing Alaska. In electing Dunleavy, Alaska voters rejected the candidacy of a Democratic challenger who opposed development of the Pebble project. Dunleavy has stated that permitting decisions about resource development proposals in Alaska, including the Pebble project, should be based on objective science and comprehensive reviews led by expert state and federal regulatory agencies.

Therefore, the final chapter of the Pebble project is yet to be written.

10

ENDANGERED SCIENCE AND THE EPA'S FINDING OF ENDANGERMENT FROM CARBON DIOXIDE

The matrix of professional incentives that distort science exists in many fields, but climate science has been especially politicized. There is essentially only one provider of climate research funding, and that is the federal government. A 2013 Office of Management and Budget report lists total "Federal Climate Change Expenditures" (including research and development) of $22.6 billion.[1] The incentive would seem strong for one important policy-related recipient of this largesse, the Environmental Protection Agency (EPA), to maintain that carbon dioxide–induced climate change is exceedingly dangerous to human health and welfare.

This was certainly a reasonable working hypothesis until the 2016 presidential election. This essay gives the scientific rationale for an EPA about-face on global warming.

History of the Endangerment Finding

In 2006, the Supreme Court granted a writ of *certiorari* to *Massachusetts v. EPA*, a case in which Massachusetts (along with 11 other states, the District of Columbia, and a plethora of environmental advocacy organizations) claimed that the Clean Air Act Amendments of 1992 contained language requiring the EPA to limit emissions of carbon dioxide from cars, because it was a "pollutant," something that endangered human health and welfare. The EPA, during the George W. Bush administration, held that this was not the case because of scientific uncertainty concerning the amount of climate change actually caused by it. In 2005, the appellate court upheld by a 2-1 vote the EPA's original decision that it did not have such authority, although the 38-page dissenting opinion by Judge David S. Tatel was impressive in its scope, legal and scientific comprehension, and length.[2]

There were serious questions, acknowledged by the Supreme Court in its majority decision to grant "cert," that the petitioners (Massachusetts and other states) might have lacked sufficient legal standing to bring the case forward, but Justice John Paul Stevens, writing for the majority, concluded that the substance of the issue trumped the legal standing of the petitioners. He wrote, "The unusual importance of the underlying issue persuaded us to grant the writ [of *certiorari*]." Subsequently, in June 2007, in a 5-4 decision, the Supreme Court held that if the EPA deemed carbon dioxide to be a pollutant that harms health and welfare, then it indeed could regulate it under the Clean Air Act. This being late in the George W. Bush administration, the EPA took a pass on regulation until after the 2008 election.

That changed about three minutes into the first Obama administration, when global warming was the second action item in his first inaugural address (after healthcare). A mere 90 days later, the EPA came out with a "Preliminary Finding of Endangerment," foreshadowing its final finding eight months later. The December 7, 2009, release date was timed to provide a U.S. climate policy bona fide for the just-started 15th Conference of the Parties to the UN's Framework Convention on Climate Change in Copenhagen. This was where the world's nations would meet to finally and definitively hammer out a new agreement to replace the failed 1997 Kyoto Protocol to reduce emissions.

The endangerment finding had to be based on some assumptions about future climate as modified by increasing atmospheric carbon dioxide, and there are really only limited tools to make this important forecast. It is not simply a matter of going back in geological time to see when atmospheric concentrations of carbon dioxide were what they might be in 2100, and then looking for proxy indicators of global temperature. By 2100, concentrations will be beyond what they were before the major glaciations that began roughly 2.6 million years ago.

The presence or absence of large areas of ice is critical to the earth's climate. Our planet in its current state, with its vast fields of polar ice, absorbs much less solar radiation than it would in their absence. There is an easy demonstration of this. On a sunny, calm day, don some light khaki trousers and sit by the pool. You'll be quite comfortable. Your outer layer has roughly the whiteness of ice. Now switch to a pair of black pants of the same material and you will get plenty hot, plenty fast. Consequently, studies that extrapolate from the past need to discount for the warming effect of ice-free surfaces that we do not enjoy now, despite the high concentration of atmospheric carbon dioxide.

One might infer cause-and-effect and say that the Greenland ice cap, as well as a substantial portion of Antarctic ice, will be lost. Maybe— but the effects are largely determined by how quickly the ice degrades. If it takes only one or two hundred years, that's catastrophic. If it takes thousands, the resultant sea-level rise will be gradual enough for easy adaptation. It is noteworthy that around 118,000 years ago, for reasons having more to do with the sun than with the atmospheric composition, Greenland's summer high-latitude temperatures averaged 6°–8°C (11°–14°F) warmer than the 20th-century average for approximately six thousand years, and it still only lost about 30 percent of its ice.[3]

This history portends a fairly resilient earth. One can extrapolate from the known changes in the radiation balance from atmospheric carbon dioxide associated with the recent glacial cycles. Even though the changes in its concentration were much smaller than what we anticipate in the future, the warming effects of carbon dioxide are known to be greatest at its lowest concentrations, so there is some legitimacy to this approach. This method tends to reduce the expected warming from prospective computer models for doubled carbon dioxide by about one-third.[4]

The scientific bases for the endangerment finding are in an accompanying Technical Support Document, which ignores such historical studies and instead relies solely on the projections of what are called general circulation models (GCMs).[5] These are complicated computer simulations of the earth's atmosphere altered by human emissions of carbon dioxide. If these can be invalidated, then so can the Technical Support Document and therefore the endangerment finding itself. By far the most cited document in the Technical Support Document is the second "National Assessment" of climate change impacts in the United States, also published in 2009. As we will see, this document and other two in its series are so flawed that they cannot seriously be considered the basis for any substantive policy.

The importance of the current endangerment finding is that it will serve as the touchstone for continual litigation of any attempt to weaken, roll back, or eliminate greenhouse gas regulations by an administration that opposes them.

The rationale for invalidation would obviously be a demonstration that the GCMs are systematically failing in their forecasts of warming. The evidence for this is in two illustrations from Dr. John Christy at the University of Alabama-Huntsville.[6] The first shows predicted and observed tropical (20°N–20°S) temperatures in the middle of the earth's active weather zone—technically the mid-troposphere, roughly from 5,000 to 30,000 feet in elevation. The predicted values are from the 102 climate model realizations from 32 different base-model groups and are given as the thin lines in Figure 10.1. These are from the most recent science compendium of the UN's Intergovernmental Panel on Climate Change (IPCC).[7] These tropics cover roughly 37.5 percent of the earth's surface.

The observations are running means of the three different histories of lower atmospheric temperatures. One such history is determined from satellite-sensed changes in the microwave emissions of oxygen, which vary with temperature. Another is the average of the four commonly used compilations of sensed temperatures from weather balloons. These should be very reliable since the instrument package is calibrated before each launch. The third is an increasingly popular "reanalysis" set of lower atmospheric temperatures derived from the initialization temperature fields from three different daily weather forecasting models. Instead of data gaps, such as occur over the Arctic

FIGURE 10.1

Tropical mid-tropospheric temperature variations, models vs. observations

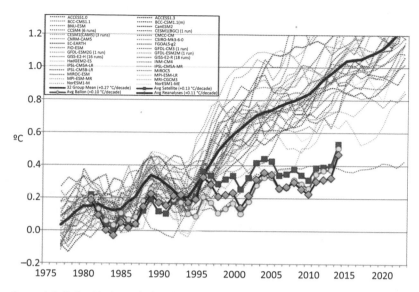

Source: J. R. Christy, Testimony before the House Committee on Science, Space and Technology, 2017, March 29, 2017, available at https://object.cato.org/sites/cato.org/files/testimony/michaels -kealey-house-testimony-march-29-2017.pdf.

Note: 5-year averages, 1979–2016. The trend line crosses zero at 1979 for all times series.

Ocean, just being infilled with extrapolation, an actual model of the physical surface is employed.

The differences between the predicted and observed changes are striking, with only one model, the Russian INM-CM4, appearing realistic. In its latest iteration, its climate sensitivity (the net warming calculated for a doubling of the atmosphere's carbon dioxide concentration) is 2.1°C (3.8°F); in comparison, the average in the family of models used in the IPCC science compendium is 3.5°C (6.3°F).[8]

Next is a somewhat more complicated illustration. It shows vertical temperatures in the tropics (see Figure 10.2). The y-axis is height, and the x-axis is temperature change since 1979, in °C/decade, predicted by

FIGURE 10.2

Tropical temperature trends, 1979–2016

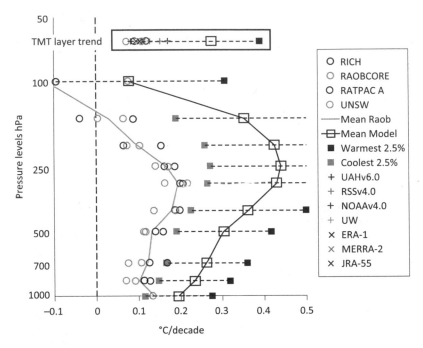

Source: J. R. Christy, Testimony before the House Committee on Science, Space and Technology, March 29, 2017, available at https://object.cato.org/sites/cato.org/files/testimony/michaels-kealey-house-testimony-march-29-2017.pdf.

Note: For 20°–20°N. Predicted (red) and observed (green) temperature trends (°C/decade) versus pressure height.

the average of the models (red) and observed from weather balloons (green). The altitude is given as the atmospheric pressure in hectopascals (hPa).[9] In reality, the altitude of different pressure surfaces varies slightly with the average temperature of the layer through which the balloon has ascended.

It is obvious that there has been a massive systematic problem with the climate models over the vast tropics since 1979. They clearly forecast a "tropical hot spot" centered from a pressure of 500 hPa (approximately 18,000 feet in altitude) to the top of the earth's active weather zone, around 150 hPa, known as the tropopause (literally, "motion stops"); in comparison, predicted warming near the surface (around 1000 hPa) is almost twice the observed value, and at the level of around 40,000 feet, the predicted rate is around *seven* times what is being observed.

The consequences of this systematic error are enormous. The vertical distribution of temperature in the tropics is central to the formation of precipitation. When the difference between the surface and the upper layers is large, surface air is more buoyant, billowing upward as the cumulonimbus cloud of a heavy thunderstorm. When the difference is less, storm activity is suppressed. According to the chart, the difference has been forecast to become much smaller than it has in reality.

Missing the tropical hot spot triggers an additional cascade of errors as a result of bad specification of precipitation. When the sun shines over a wet surface, the vast majority of its incoming energy is shunted toward the evaporation of water rather than direct heating of the surface. This is why in the hottest month of the year in Manaus, Brazil, in the middle of the tropical rainforest and only three degrees from the equator, high temperatures average only 91°F, not very different from Washington, DC's 88°F. To appreciate the effect of water on surface heating of land areas, high temperatures in July in bone-dry Death Valley average 117°F.

Getting the vertical temperature wrong will have additional consequences for precipitation. If the tropical hot spot were there, the resultant decline in precipitation, combined with very hot temperatures, could force the vegetation to change from tropical rainforest to a semi-desert.

Every person actively involved in running climate models knows all about these systematic errors (and probably many more). They know that much of the downstream "weather" resulting from an inaccurate hot spot over the entire tropics will simply be wrong, or if it is right, only fortuitously so.

We can sum up the implications of Figures 10.1 and 10.2:

ENDANGERMENT FINDING FLAW #1: **The climate models are making multiple systematic errors with regard to three-dimensional atmospheric temperatures that will dramatically affect important characteristics of climate.**

We now have enough satellite data to examine vegetation trends around the planet for the past two-plus decades. A recent comprehensive analysis was published by Zaichun Zhu and colleagues of Peking University in 2016. It contains a map of something called the "leaf area index" (LAI), which is a measure of the density and cover of vegetation. The map shows changes between 1982, when satellites first begin transmitting data, and 2009 (see Figure 10.3).

FIGURE 10.3

Trends in observed leaf area index, 1982–2009

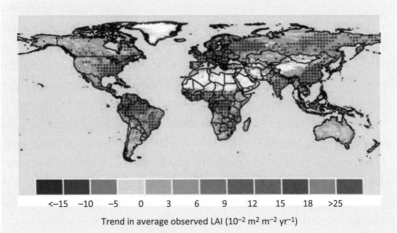

Trend in average observed LAI (10^{-2} m^2 m^{-2} yr^{-1})

Source: Z. Zhu et al., "Greening of the Earth and Its Drivers," *Nature Climate Change* 6 (2016): 791–795.

Note: In general, the earth is getting greener. In particular, the greening of the tropical rainforest areas of South America, Africa, Asia, and the large tropical islands of the western Pacific has been stunning (see the pink and magenta colors on the map).

Zhu and colleagues note that 70 percent of the planetary greening is due to the direct fertilization effect of carbon dioxide on vegetation,[10] and that CO_2 fertilization effects explain most of the greening trends in the tropics, whereas climate change resulted in greening of the high latitudes and the Tibetan Plateau.

Yes, it is true: both carbon dioxide, which is the meta-respiration of civilization, and climate change, a byproduct of increasing atmospheric concentrations, are working together to make the planet a greener, more productive place.

A more recent paper by Simon Munier, of France's Centre National de Recherches Météorologiques, and several coauthors, segregates satellite-sensed LAI data collected over the period 1999–2015 into six different vegetation types. The most common vegetation type on earth is grassland, and these researchers found a remarkable increase in LAI of 5 percent *per year*. This translates to an 85 percent increase over the 17-year model study period.[11]

Model Tuning

Left to their own devices, climate models run with increasing atmospheric carbon dioxide have long been known to produce too much warming. The tendency for carbon dioxide–driven models to overheat was explicitly recognized by the UN's Intergovernmental Panel on Climate Change (IPCC) in its 1995 "Second Assessment Report," when it stated that "most GCMs [climate models] produce a greater warming than has been observed, unless a lower climate sensitivity is used," and it claimed "growing evidence that increases in sulfate aerosols are partially counteracting the [warming].[12]

This latter hypothesis has always troubled those critical of the models because the huge uncertainty previously associated with the sulfate cooling can easily be used to "tune" the models to reproduce the climate of the 20th century. Until recently, modelers tended to elide tuning, but the magnitude of it became public in a news article in *Science* by Paul Voosen: "For years, climate scientists had been mum in public about their 'secret sauce': what happened in the models stayed in the models. The taboo reflected fears that climate contrarians would use the practice of tuning to seed doubt about [the] models."[13]

Voosen's story begins in 2010, when 32 modeling complexes were required to submit "frozen code" output to the U.S. Department of Energy so that standardized models could be compared for the upcoming 2013 IPCC scientific summary, in the fifth Climate Model

Intercomparison Project, or CMIP5. Voosen described the process for the Max Planck Institut (MPI) Model. The lead researcher, Erich Roeckner, was unavailable. So it devolved to underlings to get the model in shape to be shipped—which they could not do with ease. It appears that Roeckner alone had the expertise to tune a model's parameters in order for it to produce something that looked like a realistic climate.

They finally produced a proper model, but it projected over 7°C of warming as a result of doubled atmospheric carbon dioxide.[14] This projection would make their model by far the hottest of any that would appear in the subsequent IPCC compendium. Despite having told *Science* that "it was a damn good model," they tuned that large warming away by adjusting other parameters, such as the dispersal of heat through the ocean. Finally, after many months, MPI had a product for the CMIP.

However, for the first time in recent memory, Roeckner's group decided to document what they had done. Thorsten Mauritsen, another of MPI's senior scientists, published the 2012 paper in the *Journal of Advances in Modeling Earth Systems*, noting that it was hardly the definitive encyclopedia of tuning, because it is apparently impossible to know what was done to the models over their historical development. In Mauritsen's words, "Model development happens over generations, and it is difficult to describe comprehensively."[15]

That's because so many of the people who work on these models are temporary or ephemeral, like graduate students and postdoctoral fellows, and they don't always leave notes about what kinds of tuning they did. In fact, they usually don't. Significant portions of climate models are therefore black boxes with varying degrees of subjectivity. It is the subjective modeler and not the objective model that determines future climate.

Indeed, in the article by Mauritsen and colleagues we find a tremendous number of tuned parameters. It seems telling that although almost all models are tuned to replicate known global temperatures of the 20th century, the range of model sensitivity is on the order of several degrees Celsius. Voosen described the process for the Max Planck Institute Model: "MPIM hadn't tuned for sensitivity before—it was a point of pride—but they had to get that number down."

The news hook for Voosen's article was a landmark paper by Frederic Hourdin, (the head of the French modeling effort), Mauritsen, and

13 other coauthors from different modeling groups that had been accepted for publication in the *Bulletin of the American Meteorological Society*. The paper had the cheeky title, "The Art and Science of Model Tuning." Hourdin and his colleagues described the process of tuning:

> One can imagine changing a parameter which is known to affect the sensitivity, keeping both this parameter and the ECS [equilibrium climate sensitivity] in the *anticipated acceptable range* [emphasis added].[16]

Voosen went on to note that all climate models are tuned to simulate the climate history of the 20th century.[17] In a remarkable passage, Hourdin and his coauthors wrote:

> In fact, the tuning strategy was not even part of the required documentation in the CMIP5 simulations. . . . Why such a lack of transparency? Maybe because tuning is often seen as an unavoidable but dirty part of climate modeling. Some may be concerned that explaining that models are tuned could strengthen the arguments of those who question the validity of climate change projections.[18]

On her popular blog, *Climate Etc.* (www.judithcurry.com), Judith Curry, a now-retired Georgia Tech climate scientist, wrote about the paper by Hourdin and others: "If ever in your life you are to read one paper on climate modeling, this is the paper that you should read."[19]

The problem with model tuning is that code is changed in ways that may not be physically realistic if the objective is for a match with the 20th-century global temperature history. As a result, these same alterations, which now exist in "frozen code" for the IPCC climate compendium, make their 21st-century predictions with parameters that in some cases are simply not correct. This is a very plausible explanation for their massive departures from the reality of both horizontal and vertical temperatures (see Figures 10.1 and 10.2).

The core claim that will be used against the endangerment finding, which is solely based on these models for future climate, is this: rather than the physics of the model predicting future warming, it is the modeler who will ultimately choose what warming is scientifically acceptable. Tuning the climate model matters because there are so many tunable parameters, and the range of possible parameter values can be so large as to allow any result.

ENDANGERMENT FINDING FLAW #2: **Climate models are "tuned" to produce what is a subjectively determined "acceptable" amount of climate change. They are simply not mature enough to be used as the basis for expansive policies.**

DID THE 2016 ELECTION TRUMP THE HOURDIN PAPER?

It is fair to ask what prompted the publication of Hourdin and his colleagues' candid manuscript on model tuning. According to the *Bulletin of the American Meteorological Society*, it was received in final form on July 9, 2016, and published online on March 17, 2017. Clearly the manuscript must have been in preparation for much of 2015. If submitted around January 2016, the date of the final manuscript suggests it was probably subject to two revisions at the suggestion of reviewers.

Copies (including mine) were circulating everywhere before official publication. Judith Curry's oft-cited blog post was published on August 1, 2016, and Voosen's widely read *Science* revelation appeared on October 28, 2016. He interviewed the authors of the paper but never mentioned the manuscript itself, obviously to keep *Science* in the good graces of the American Meteorological Society by not scooping it.

As noted by Voosen, modelers worried that revealing the extent and the subjectivity of the tuning process could jeopardize climate policy. But that concern would be valid only if there were also a president (Trump) who wanted to change policy, which would have seemed a very long shot when the paper was submitted. So, under all but the most improbable eventuality, it was a very strategic move to open up about tuning and its associated problems, with the hope of securing more research funding from an incoming Clinton administration.

As already noted, the Technical Support Document relies heavily on the second of three serial documents put out by the U.S. Global Change Research Program (USGCRP). The Global Change Research Act of 1990 created the USGCRP and mandates that it produce periodic assessments of the effects of global climate change on the United States. These are the "National Assessments," the first of which appeared in November 2000, after Election Day (but not after the election was settled) and before the inauguration of George W. Bush.

The National Assessments are compendia of the putative effects of domestic climate change that form the basis for U.S. policy. But all of the assessments have been seriously flawed and appear more as polemic than as objective and dispassionate analyses. They base their prospective climate impacts solely on the climate models.

The first assessment used the temperature and precipitation output of two climate models to drive "effects" models on various sectors, such as agriculture, forestry, and human health. The USGCRP Synthesis Team, headed by Tom Karl, then director of the National Climatic Data Center in Asheville, North Carolina, had nine such models to choose from, and the two it chose were from the Canadian Climate Centre and Britain's Hadley Center, a component of the United Kingdom Meteorological Office that specializes in climate modeling.

The Canadian model produced the largest 21st-century temperature changes of any of the nine models, and the Hadley version produced the largest precipitation changes. When I asked the director of the USGCRP, Mike MacCracken, about this, he replied that they "wanted to look at the most extreme possibilities."

"NEGATIVE KNOWLEDGE"

Every scientific "model" is actually a hypothesis about the way a system behaves, and hypotheses need to be tested with real-world data to see if they can continue to be entertained, or if they need to be modified.

In the case of a climate model, the hypothesis is that the model is accurately simulating temperature changes in the real atmosphere as carbon dioxide rises. If all models are tuned to be able to mimic the global temperature history of the 20th century, this should seem like a cinch, right?

Except, as noted above, the global models used in the (first) 2000 National Assessment were chosen because they simulated the largest changes in temperature (the Canadian model) and precipitation (the Hadley model) of the nine considered models. Could that mean that they might be exaggerating observed climate change in the United States?

In an attempt to answer this, Paul C. Knappenberger and I examined the predicted and observed 10-year running means of U.S. temperatures—that is, 1901–1910, 1902–1911, etc. We first looked at the period-to-period variability of the raw data. If a model is working and we apply it to those data, what's left over (that is, what is *not* explained by the model) will have a variability that is significantly less than the raw data. In other words, the model will "explain" a portion of the variability of the raw data. If, however, somehow the model-minus-observed data variability is *greater* than that of the raw data, the model has failed so badly that it has actually added negative knowledge.

This is no different than a student scoring less than 25 percent on a four-option multiple-choice exam. It means that his or her synthesis of the subject matter is somehow worse than it was before taking the course. That is precisely the behavior of the two climate models that underpinned the first National Assessment. Both the Canadian and Hadley models *added* variability to the raw data when they were applied.[20]

Now that you've read the sidebar "Negative Knowledge," you know how well the first National Assessment models worked. They didn't.

FIGURE 10.4

Observed and modeled 20th-century temperature changes

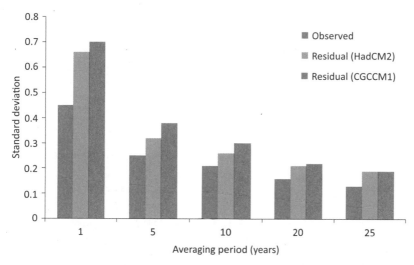

Source: Original data supplied in personal communication from Tom Karl, 2000.

Note: blue = raw U.S. temperature data variability; green = variability after the Hadley model was applied; gray = variability after the Canadian model was applied. For both models, and at all time intervals, the models increased the variability of the data after they were applied.

So, after completing my peer review, I reported my findings to Karl. I explained to him that using these models to predict the effects of climate change on the United States is analogous to a physician prescribing a medication to a patient that will have the opposite of the desired effect. It would be like prescribing adrenaline for high blood pressure, and that would be called malpractice.

Impressively, Karl emailed back. His team had applied not only my ten-year test to the models, but also to one, five, twenty, and twenty-five year running means. *At all time frames the models added variability to the raw data* instead of reducing it. The relevant illustration is in Figure 10.4.

The second National Assessment came out soon after the end of the George W. Bush administration, in June 2009. It was obviously produced almost exclusively during his administration, as such reports take years to develop. Our review, dated August 2008, began with this paragraph:

Of all of the "consensus" government or intergovernmental documents of this genre that [we] have reviewed in [our] 30+ years in this profession, there is no doubt that this is the absolute worst of all. Virtually every sentence can be contested or does not represent a complete survey of a relevant literature.

To prove our point, I assembled a six-person team to produce a palimpsest called *Addendum: Global Climate Impacts in the United States*. It was entirely analogous to the 2009 federal report. For example, under "Key Findings," the government version says:

7. Risks to human health will increase. Harmful health impacts are related to increasing heat stress, waterborne diseases, poor air quality, extreme weather events, and diseases transmitted by insects and rodents. Reduced cold stress provides some benefits. Robust public health infrastructure can reduce the potential for negative impacts. (p. 89)[21]

Our palimpsest says:

7. Life expectancy and wealth are likely to continue to increase. There is little relationship between climate and life expectancy and wealth. Even under the most dire climate scenarios, people will be much healthier and wealthier in the year 2100 than they are today.[22]

The *Addendum* was especially richly referenced in the sections on agriculture and ecosystems, two fields that have a much more balanced literature than, say, climate science. The reasons for this are documented in the 2016 book *Lukewarming*.[23] In part because of that substantial literature, *Addendum* had nearly twice as many scientific citations as the second National Assessment. Karl was also the lead author of the second assessment.[24]

The second assessment was *the* critical component of the Technical Support Document for the EPA's 2009 endangerment finding.[25]

Its prospective forecasts on all of the impact areas (agriculture, human health, etc.) and regions of the United States are all based on models that have massive systematic errors. These include an incorrect simulation of the evolution of tropical lower atmospheric temperatures since 1979 that erroneously projects an upper-atmospheric "hot spot" over the entire tropics.

Further, the 2009 National Assessment necessarily summarized a literature heavily biased by the incentive structure in modern science, as detailed in this volume's introductory chapters.

ENDANGERMENT FINDING FLAW #3: **The Technical Support Document for the endangerment finding is largely based on the 2009 National Assessment, which is itself based on models with large systematic errors and a literature demonstrably biased toward dire climate findings.**

Five years later, the third National Assessment was published. This one was blatantly couched in the context of the Obama administration's activism on climate change. In fact, in the May 6, 2014, introduction, the National Oceanic and Atmospheric Administration wrote:

> The report, *a key deliverable of President Obama's Climate Action Plan*, is the most comprehensive and authoritative scientific report ever generated about climate changes that are happening now in the United States and further changes that we can expect to see throughout this century. [emphasis added][26]

All three National Assessments are deeply flawed. The first broke the normative rule that models should be used for policy only if they have explanatory capability. The second was so incomplete that it could be fodder for an entire palimpsest with nearly twice as many refereed citations, and the third was specifically designed as a part of the Obama administration's policy thrusts on climate change.[27]

In summary, the endangerment finding is extremely vulnerable. It is based on models that dramatically misrepresent the tropical lower atmosphere's evolution over the past 38 years (that is, since the beginning of the satellite-sensed temperature record). It is based on models that are subjectively tuned to an "anticipated acceptable range" of output while simulating the 20th century's climate. And it is largely based on systematically flawed National Assessments of climate change impacts on the United States.

The endangerment finding can and should be vacated because it is based on models that simply do not work.

THE EPA'S CONFLICTED "SCIENCE" ON FINE
PARTICULATE MORTALITY

n this book are several examples of "science by authority" in service of increasing regulation or putting up regulatory roadblocks. Panels of accomplished scientists are asked by government agencies to summarize the state of science on some important or politically hot topic. The people who sit on such panels have often received massive public funding to study what they are empaneled to summarize. One prominent example is found in panels appointed by the U.S. Environmental Protection Agency's (EPA) Clean Air Scientific Advisory Committee (CASAC). Appointed to review the scientific basis for EPA regulations, the members of such panels typically review their own EPA-funded work.

On October 31, 2017, EPA administrator Scott Pruitt announced that members of the agency's multiple science advisory boards and related administrative structures would no longer be eligible to receive research funding from his agency. "It is very, very important to ensure independence, to ensure that we're getting advice and counsel independent of the EPA," he noted. As an example, Pruitt said that scientists on only three such boards had received $77 *million* in the previous three years.

Seemingly turning logic on its head, Andrew Rosenberg of the Union of Concerned Scientists told the *Washington Post*'s Juliet Eilperin that, on the contrary, "The consequences of [the new EPA rule] aren't just bad for a few scientists. This could mean that there's no independent voice ensuring that the EPA follows the science on everything from drinking water pollution to atmospheric chemical exposure.[1] Rosenberg's concept of independence is examined closely in this chapter. In reality, Pruitt had identified a major problem at the EPA. In many cases, the agency (and many others like it) relied on the scientists that it funded for advice and policy prescriptions related to its funded scientists' research.

This chapter takes a deep dive into how this can create a massive regulatory undertaking based on questionable or omitted science. Here we look at the regulatory process that set the limits for particulate matter of less than 2.5 micrometers in diameter, known as $PM_{2.5}$, or fine PM. Fine PM is ubiquitous in the air we breathe and is generated by a myriad of human activities, including such primary industries as agriculture and mining (see Figure 11.1) and also everyday actions such as driving a diesel car.

In 2015, the Obama administration's Office of Management and Budget (OMB) claimed that the major air pollution regulations in the previous decade generated between $158 and $778 billion in benefits at a cost of only $37 to $44 billion.[2] This is 99 percent of the total benefits from all EPA regulations and an astounding 60–80 percent of the benefits of *all* federal regulations issued between 2004 and 2014.

But most of these massive benefits accrue from only one regulatory target: fine PM.

As can be seen from Figure 11.2, which is based on a 2013 OMB report, the argument that total federal regulations over the past fifteen years are cost-benefit justified largely depends on whether or not the benefits of lowering levels of fine PM are as big as the EPA says.[3]

The EPA derives the benefits from lowering (toughening) the fine PM standard by multiplying its estimate of the lives saved by a dollar value of statistical life saved. Economists supply the value of a statistical life, but it is the EPA's assessment of the underlying "science" that provides the number of statistical lives.

FIGURE 11.1

An Australian open-pit mine

Source: Quentin Jones, Sydney Morning Herald, Fairfax Syndication.

Note: Fine and coarse particulates are everywhere. A large tilling or harvesting rig in modern agriculture generates a similar cloud.

Some hypothetical (but reasonable) numbers will clarify how the EPA does this. Suppose the agency concludes that there will be a 1 percent decrease in annual mortality in a given population if the fine particulate standard is reduced to, say, 10 μg/m³ (this is the standard unit of micrograms per cubic meter). If there are 100,000 people in the population exposed to fine particulates, then the EPA would conclude that there are 1,000 statistical lives saved per year by reducing the fine PM standard to 10. Economists have come up with estimates of the dollar value of a statistical life of around $6 million. Multiplying $6 million by the 1,000 statistical lives saved annually, the EPA would find the benefit of the regulation to be $6 billion.

What is the "science" behind this determination, and how was it assessed and reviewed by the EPA? That is what I examine in this chapter,

FIGURE 11.2
Benefits from fine PM reduction make up most of the OMB's monetized regulatory benefits

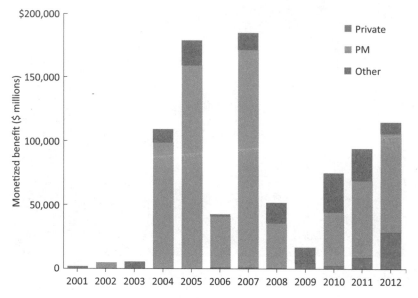

Source: Reprinted from Susan E. Dudley, "OMB's Reported Benefits of Regulation: Too Good to Be True?" *Regulation* 36 (2013): Figure 1.

by taking a critical look at the statistical work that the EPA claims shows that decreasing fine PM levels led to discrete reductions in mortality. The agency's 2009 scientific assessment of fine PM supported its tightening of the fine PM annual average standard to 12 µg/m³ (from 15 µg/m³) in 2012.[4] In making the ultimate regulatory recommendation, the EPA completely ignored not only obvious problems with studies that it claimed to show causation, but also an extensive literature that explored methodological failures of those studies and that investigated alternative mortality influences.

The fine PM science was a product of the CASAC empaneled to advise on fine PM. According to the Clean Air Act of 1970 (as amended in 1977), the CASAC is supposed to provide an "independent" scientific review of the agency's scientific assessments. However, this panel consisted entirely of researchers working at EPA-funded academic centers producing research showing that fine PM causes increases in mortality. The "independent" review panel thus consisted of scientists who were heavily dependent on EPA funding. This panel makeup is flatly inconsistent with Congress's intent in establishing such scientific review boards.

The EPA's fine particulate review panels are like a group of students grading their own papers. By putting the producers of the research in charge of reviewing it, the EPA guaranteed that its fine PM assessment will never by objectively critiqued. Here, the EPA became a vertical conglomerate, dominating the research, assessments, assessment reviews, and regulatory outcome.

How the EPA Sets Air Quality Standards

Under Section 108 of the Clean Air Act (CAA) of 1970, the EPA is to set "primary" or health-based, national ambient air quality standards (often abbreviated NAAQS) at levels that are "requisite to protect the public health . . . allowing for an adequate margin of safety,"[5] based on "air quality criteria [that] shall accurately reflect the latest scientific knowledge useful in indicating the kind and extent of all identifiable effects on public health or welfare which may be expected from the presence of such pollutant in the ambient air, in varying quantities."[6] The 1977 amendments to the CAA[7] required that, at least every five years, the EPA "shall complete a thorough review of the criteria . . . and the national ambient air quality standards promulgated under this section and shall make such revisions in such criteria and standards and promulgate such new standards as may be appropriate."[8]

Setting Science-Based National Standards

The key step in the EPA's five-year reviews of the national air quality standards is called an Integrated Science Assessment (ISA). In the ISA, EPA staffers review what may be thousands of published studies "to draw conclusions regarding the causal relationships between exposure to each criteria pollutant and human health and welfare effects."[9] Criteria pollutants are the conventional air pollutants specifically covered by the Clean Air Act, such as carbon monoxide. As the decision facing the EPA is typically whether to retain or instead raise (tighten) a standard for allowable concentration of a criteria air pollutant, what the ISA says about the causal effect on human health of reducing concentration levels at the level of the current standard carries "substantial weight in the Administrator's judgments."[10]

The EPA states that in producing an ISA, "all relevant controlled human exposure, epidemiology and toxicology studies are considered."[11] Moreover, for the EPA to conclude that the evidence is "inadequate to infer a causal relationship," there must *not even be a single study finding such a relationship*. Indeed, if there is a "single high-equality epidemiologic study [that] shows an association with a given health outcome but the results of other studies are inconsistent," then the EPA may conclude that the existing literature is "suggestive of a causal relationship." If the "pollutant has been shown to result in health effects in studies in which chance, bias, and confounding could be ruled out with reasonable confidence," then the EPA may reach its strongest causal determination, that there is a causal relationship between exposure at the current standard and adverse human health effects.[12]

As noted by Julie E. Goodman and coauthors, there are several problems with the EPA's framework for determining the causal effect of further reductions in air pollution:

> There is no explicit guidance for evaluating the strengths and weaknesses of studies and no information on how to use study quality criteria to assign a quality weight to individual studies . . . there is no indication of what constitutes a strong association. . . . There is also no clear guidance on how to evaluate alternative hypotheses that are supported by the data (e.g. a given substance is not a causal factor for a health effect and apparent associations are explicable by other factors) or what criteria should be used to determine which hypothesis is most supported by the data.[13]

This lack of guidance will lead to the ad hoc science that I describe below.

After the production of the ISA, EPA staff use the findings of the ISA to produce a shorter document, a risk exposure assessment

(REA). In the REA, the staff interprets the ISA to "generate a single concentration-response model to predict health effects at different levels of exposure."[14] After the REA is completed, EPA staff prepare a Policy Assessment in which they attempt to connect the "scientific assessments in the ISA and REA and judgments required by the EPA Administrator" regarding any potential revision of a NAAQS.[15] Throughout this process, EPA staff is advised by the Clean Air Act Science Advisory Committee. Under the 1977 amendments to the CAA, Congress required that CASAC be an "independent scientific review committee."[16] CASAC review occurs at virtually every stage of the process by which EPA staff develop proposals for new NAAQS: when staff formulate policy issues and scientific questions, in the development of both the ISA and the risk exposure assessment, and before the presentation of the final policy assessment to the EPA administrator.

Setting the December 2012 New Fine PM Standard

In December 2012, the EPA promulgated its new fine particulate standard, lowering the allowable annual average ambient level from 15 μg/m^3 to 12 μg/m^3. The process leading to that new standard began in 2007, when EPA staff began combing the literature for studies published since its previous ISA on fine particulate matter in 2002. As EPA staff explain in the final product of this review, the December 2009 ISA, "much of the newly available health information evaluated in this ISA comes from epidemiologic studies that report a statistical association between ambient exposure and health outcomes."[17] Of those health outcomes, by far the most important in the EPA's cost-benefit analysis of the regulation was a reduction in mortality.[18]

Strikingly, the resultant Regulatory Impact Analysis (RIA) cited only two studies in its estimates of the monetized benefits of reduced premature mortality.[19] The EPA's cost-benefit analysis takes those studies as estimating that between 460 and 1,000 premature deaths per year, valued at between $4 billion and $9 billion, will be saved by the new fine PM standard. But those studies in fact support a further reduction in the fine PM standard, for they show that between 1,500 and 3,300 premature deaths, valued at between $13 and $29 billion, would be avoided every year by lowering the fine PM standard all the way to 11 µg/m³.

According to the RIA, the change in annual benefits (marginal benefit) that would be achieved by dropping the standard from 13 to 12 is between $2.7 and $6.1 billion. But the marginal benefit in going even further, from 12 to 11, would be between $9 and $20 billion. The EPA estimates that the marginal cost of moving to 12 would be at most $350 million, and that of moving to 11 would be at most $1.7 billion. On these numbers, moving to 11 would generate net benefits of at least $7.3 billion. On basic cost-benefit principles, the EPA's 2012 RIA for the fine PM standard makes the case for going at least to 11 a no-brainer. The fact that this stricter standard was not adopted suggests that even the EPA does not believe its own numbers; they would appear to have been generated *only* in order to reach a predetermined result.

Given that the benefits from EPA's 2012 reduction in the fine PM standard are the vast majority of all quantified benefits of EPA regulations since 2004, and given that those benefits flow almost entirely from the supposed reduction in premature mortality caused by the reduction in fine PM, it is important to take a closer look at the epidemiological studies in the EPA's ISA that establish what the EPA's own CASAC PM review panel says show a "causal" or "likely causal" relationship between reductions in fine PM below 15 µg/m³ and reductions in mortality.

The EPA's Evidence for the 2012 Fine PM Standard

The primary basis for the EPA's estimation of the health benefits from reducing the standard from 15 to 12 µg/m³ are epidemiological studies of different populations of people exposed to different levels of fine PM.

Also called observational studies, they are to be distinguished from toxicological studies in which rats or mice are exposed to unrealistically high levels of a pollutant, and also from the randomized clinical trials used in drug research. Of these three, observational studies are the least rigorous, and the greater prevalence of confounding factors in observational studies than in clinical trials is obvious, in both short-term and especially long-term studies.

Short-Term Observational Studies

In a short-term study, the investigator gathers data on mortality and fine PM levels for a given set of localities (often a single city or county) and then asks whether fine PM levels on a given day are statistically related to mortality rates within a window of some number of subsequent days. If the finding is that there is a statistically significant positive relationship, meaning that days with higher measured levels of fine PM are followed by a day or days with higher mortality rates, then this is taken as evidence that increases in fine PM cause increases in mortality.

In its fine PM science assessment, the EPA stressed short-term mortality studies that pool many cities together. According to the EPA, the estimated increase in mortality risk due to fine PM exposure from such studies ranges from .29 to 1.21 percent.[20]

THE EPA'S SUMMARY VERSUS WHAT SHORT-TERM
TIME SERIES ACTUALLY FIND

To better understand not only what these numbers mean, but also what they conceal, consider in a bit more detail the study with the highest estimated impact of fine PM exposure on mortality.[21] In that study, Meredith Franklin and coauthors investigated whether the difference in deaths between any day in a month and deaths on the third day of the month was statistically related to the difference in fine PM levels between the two days.[22] They considered 27 cities that had at least two years of data on fine PM levels over the period from January 1, 1997, to December 31, 2002, and found that mortality from all causes increased by 1.21 percent for every 10 $\mu g/m^3$ increase in fine PM.

To put this in concrete numbers, consider one of the 27 cities studied, Milwaukee. According to Franklin and coauthors, there were 36,637 deaths in Milwaukee over the four years studied, 1997–2000.[23] Neglecting seasonal variation just for purposes of this example, this works out to an average of 25 deaths per day. An increase of 1.2 percent from this base of 25 is equal to 0.3 additional deaths per day on average. Therefore, if Milwaukee had experienced the average impact (over all 27 cities) from an increase of 10 $\mu g/m^3$ in fine PM, then there would have been an additional nine deaths per month, or about 110 deaths per year in that city due to fine PM exposure.

This seems like a very high number, but the actual relationship between fine PM exposure and mortality estimated for Milwaukee by Franklin and coauthors is even bigger. Indeed, in Milwaukee fine PM had a greater impact on increasing mortality than any city they studied. This should have been a big red flag to regulators. In five cities (Birmingham, Dallas, Houston, Riverside, and Las Vegas [marginally]) the error bars for a negative relationship did not overlap with zero, so increases in fine PM exposure actually *decreased* mortality. The EPA's assessment acknowledges this very surprising finding of a negative relationship, but downplays it, dismissing it with the assertion that although "it is unclear why these cities exhibit negative (and significant) risk estimates rather than null effects," "most" of the cities with a negative relationship between fine PM levels and mortality have a "high prevalence of central air conditioning."[24]

In this comment, the EPA's assessment conveys a misleading picture of what Franklin and coauthors actually found. They did not find a negative relationship between fine PM exposure and mortality just in cities with a high proportion of homes with air conditioning. Indeed, 17 of the 27 cities had error bars that included a negative relationship (see Figure 11.3).

From Figure 11.3, one can see that Milwaukee and Phoenix are outliers, the only cities where an increase of up to 10 percent in mortality cannot be ruled out using standard assumptions about the statistical distribution of observations. The closeness of the "summary" data to the zero line indicates that if Milwaukee and Phoenix were removed from the study, the relationship between fine PM and mortality would probably be statistically indistinguishable from *no relationship at all*. From these data alone, one would infer that fine PM is probably relatively unimportant in determining daily mortality spikes within a metropolitan region.[25]

There is another important feature of the results found by Franklin and coauthors that is entirely neglected in the EPA's fine PM assessment. The statistical significance accrued only for people aged 75 or older, a population with an 11-year life expectancy; in comparison, a

FIGURE 11.3

Percentage change in all-cause mortality associated with a 10 µg/m³ increase in the previous day's fine PM concentration

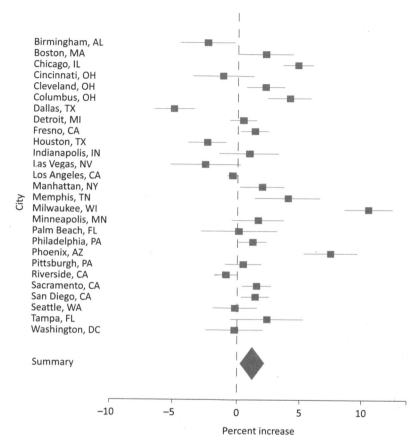

Source: Meredith Franklin, Ariana Zeka, and Joel Schwartz, "Association between PM₂.₅ and All-Cause and Specific-Cause Mortality in 27 U.S. Communities," *Journal of Exposure Science and Environmental Epidemiology* 17 (2007): 283.

Note: Increases are relative to the control measure date of the third day of the month. The shaded lines give the 95 percent confidence bounds.

30-year-old has 48 expected additional years of life.[26] Whatever the real value of a life may be, the cost of increases in mortality estimated to flow from increases in fine PM—in terms of lost expected life years—is much lower for a 75-year-old than for a 30-year-old. The EPA failed to take account of this important feature of mortality costs when calculating the benefits from reducing fine PM.

SHORT-TERM TIME-SERIES STUDIES ALWAYS FIND THAT
INCREASES IN FINE PM DECREASE MORTALITY IN MANY CITIES

The findings by Franklin and coauthors of some significant negative relationships are well known among environmental epidemiologists. In another short-term study relied on by the EPA in its 2009 fine PM assessment, Bart Ostro and coauthors found that for six of the nine California counties examined, there was at least one negative relationship between increases in fine PM and cause-specific mortality. For Kern, Los Angeles, Riverside, and Santa Clara counties (almost half of those studied), Ostro and coauthors found that mortality among people over 65 years of age decreased as $PM_{2.5}$ increased.[27]

In 2017, a group of EPA researchers themselves called the "inability to explain" the substantial variation in the relationship between fine PM exposure and mortality observed across different cities a "key uncertainty."[28] Figure 11.4, taken from that publication, clearly depicts the problem. This study found that in only four of the 77 cities studied—Fresno, Greenville, Oklahoma City, and Raleigh—could one statistically rule out a negative relationship between increases in fine PM and mortality (at the standard 95 percent confidence level).

Long-Term Cohort Studies

The EPA also relied on long-term studies of large cohorts of individuals. At enrollment, subjects were asked about personal risk factors, smoking, occupation, and age; then mortality and pollution levels were tracked over the following years. Data were also gathered on characteristics of the subjects' communities, such as average income and poverty rate. With fine PM measured at varying intervals over the study period, using varying statistical models, researchers estimate the relationship between fine PM levels and mortality rates.

FIGURE 11.4

Variation across cities in impact of PM$_{2.5}$ on mortality

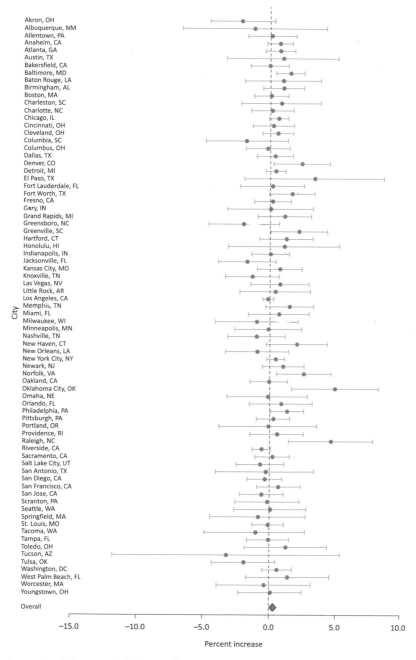

Source: Lisa K. Baxter et al., "Influence of Exposure Differences on City-to-City Heterogeneity in PM$_{2.5}$-Mortality Associations in U.S. Cities," *Environmental Health* 16 (2017): 1–8.

Note: Percent increase in non-accidental mortality with a 10 μg/m3 increase in PM2.5 concentrations using a moving average of the previous 2-days (lag 0–1) (circles = estimate; whiskers = 95% confidence interval).

Only one study, by Daniel Krewski and coauthors, was available for the 2009 assessment; it is often called the American Cancer Society (ACS) study.[29] For the subsequent RIA, another study, by Johanna Lepuele and coauthors, was available as well, reporting results on what is known as the Harvard Six Cities study.[30]

Lepuele and coauthors gathered data on a random sample of 8,111 white people from six disparate cities who ranged in age from 25 through 74 at enrollment. The data included background personal information and medical history. In their 2009 study, Krewski and coauthors investigated the prospective mortality of 1.2 million adults over age 30 (and members of households with at least one person older than 45) residing in 151 continental U.S. cities, with air pollution, lifestyle, and medical history data included, as well as zip-code-socioeconomic data. They found that a 10 $\mu g/m^3$ increase in fine PM leads to an expected total (or all-cause) mortality increase in the range of 3–4 percent. They also presented a finer-grained analysis of both Los Angeles and New York. Those analyses suggest that the national analysis means much less than the EPA says.

Krewski and coauthors estimated the fine PM concentration at the zip code level for both New York and Los Angeles. In New York, they found that for the mortality risks one expects to be most affected by pollution—all-cause, cardiopulmonary, and lung cancer deaths—the data were extremely noisy, with the aggregate error bars occupying more space on the negative (mortality falling) side of the ledger, as fine PM concentration increases.

These results are displayed in Figure 11.5. It shows the relative risk (the hazard ratio) of various types of mortality associated with an increase of 1.5 $\mu g/m^3$ exposure to $PM_{2.5}$.[31] Note that a hazard ratio below 1.0 means that exposure was associated with a lower mortality level. As the bars give the 95 percent confidence intervals, it can be seen from Figure 11.4 that the only New York mortality risk for which a zero or negative relationship to $PM_{2.5}$ could be statistically ruled out was ischemic heart disease (IHD) mortality in the winter.[32]

In Los Angeles, Krewski and coauthors did consistently find that a 10 $\mu g/m^3$ increase in estimated zip code level fine PM was associated with an increase in all-cause mortality and a slightly larger increase

FIGURE 11.5

Mortality and fine PM exposure relationship for the New York City area, 2000–2002

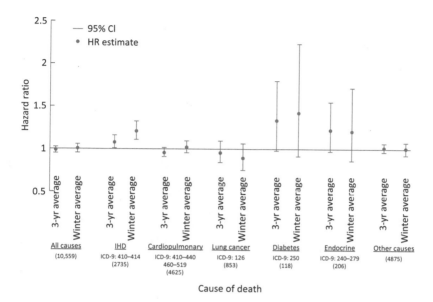

Source: Louis Anthony Cox and Douglas A. Popken, "Has Reducing Fine Particulate Matter and Ozone Reduced Mortality Rates in the United States?" *Annals of Epidemiology* 25 (2015): 53.

in IHD-related mortality. On the other hand, even for the mortality causes with statistically significant risk increases (all-cause, IHD, and cardiopulmonary death), only for all-cause and IHD mortality did the 95 percent confidence interval rule out an actual decrease in risk.

What is one to make of the starkly different associations between fine PM and mortality risk—with a generally negative relationship in New York, but a positive relationship in Los Angeles? Many Los Angeles zip codes are socioeconomically heterogeneous.[33] Consequently, for Los Angeles, the zip-code-socioeconomic variable used by Krewski and coauthors was an extremely noisy and error-prone measure of socioeconomic variables at the level that matters, the neighborhood. Similarly, the linear regression model that they used to estimate zip code level $PM_{2.5}$ was wildly inaccurate for Los Angeles, vastly overpredicting fine PM concentrations at monitors located near intersections of major freeways, with some overpredictions exceeding 126 and many exceeding 50 $\mu g/m^3$.[34]

In the language of statistics, both of these explanatory variables suffer from measurement error. Under measurement error, the estimated relationships between mortality and fine PM and mortality and neighborhood socioeconomic status (SES) are biased.

In the New York area, by contrast, the actual three-year average fine PM (between 10 and 20 $\mu g/m^3$) was in the same range as predicted by Krewski and coauthors' linear land use regression model. Moreover, the relationship between zip codes and neighborhoods in New York City is the opposite of that in Los Angeles: no zip code in New York City overlaps neighborhoods; rather, a single neighborhood may have several zip codes.[35] One would consequently expect the SES to influence their estimated impact of fine PM, which is what Krewski and coauthors found.[36] Indeed, for New York, controlling for SES measures made the association of fine PM with mortality more negative for every cause except ischemic heart disease.

The EPA Assessment Finds That Fine PM Increases Mortality Despite Evidence to the Contrary

With studies finding such noisy relationships between fine PM and mortality for so many cities, one may well wonder how the EPA's assessment of those studies could conclude that increases in fine PM actually increase mortality. Instead of going back to more carefully measure neighborhood characteristics and the response to fine PM, the EPA-funded researchers instead rely on bigger and bigger statistical samples and particular statistical models. With a big enough sample and the "right" statistical model, studies are able to report a positive and statistically significant relationship between fine PM and mortality.[37]

The Cost of Statistical Precision

The EPA clearly believes that the strongest evidence for the causal effect of exposure to fine PM comes from very large population studies, especially the 2009 study by Krewski and coauthors, with over

1 million participants. There is a simple reason for this. As the size of the sample in a statistical study increases, almost any estimated effect, however small, becomes statistically significant.[38] The very large studies to which the EPA attaches determinative weight in its fine PM assessment do find many statistically significant effects of fine PM. But just as consistent is the finding of statistically significant but negative effects of fine PM exposure on mortality.

The persistence of this problem, what Ostro and coauthors call "heterogeneity in the findings,"[39] is the cost of devoting resources to more and more analyses of large studies rather than to the construction of new and better data sets that might reveal what is actually going on. Both the short- and long-term studies that the EPA's fine PM risk assessment sets out as establishing the case for the adverse impact of increases in fine PM suffer from a basic problem: they look at such large geographic areas that they are unable to track the actual exposure of people to fine PM (using instead highly error-prone and sometimes biased estimates of actual exposure levels in the geographic unit studied); even worse, they are unable to control for the large number of neighborhood-level and individual features that are known to be primary determinants of disease and mortality.[40] These problems have beset virtually all the studies on the supposed causal link between fine PM exposure and various mortality measures.

Similar but even worse problems are present in the long-term cohort studies. Krewski and coauthors say the spatial relationship between their socioeconomic variables (what they call ecological covariates) and exposure to fine PM is "complex," but as noted above, a more pointed description might be that their measures are likely highly error prone and random.[41]

The study by Krewski and coauthors that the EPA assessment report relied on itself shows the importance of precisely measuring and properly controlling for lifestyle and socioeconomic factors. After all, they found that in New York, "pollution appeared worst where people seemed to be healthier and wealthier . . . pollution is less likely to have a major impact on health because it must compete with many positive health attributes in the individuals such as good nutrition, clean employment, and access to medical care." For New

York, Krewski and coauthors had something closer to a real neighborhood measure of socioeconomic factors (because in New York the geographic unit used, the zip code, is smaller than a neighborhood). What they are saying in the quotation above is that higher levels of $PM_{2.5}$ had little if any statistically significant health impact in zip codes where people were healthier and wealthier. This suggests that socioeconomic factors, not the level of fine $PM_{2.5}$, actually explain mortality.

Recent work shows, contrary to the implicit assumption of virtually all of the work relied upon by the EPA in its 2009 fine particulate risk assessment, that neighborhoods within a city or county are exposed to very different types of fine particulates and at very different concentration levels. A 2016 study, for example, finds up to a 20-fold variation within New York City in exposure to fine PM emissions from vehicular traffic (the major source of fine PM in the city). High-poverty neighborhoods experienced the highest levels of exposure.[42]

Model Selection Bias

Another equally important problem with the studies that the EPA relies on is model selection bias. This bias arises when the size and statistical significance of the estimated impact of the explanatory variable of key interest (such as fine PM) on the dependent variable (such as the daily mortality rate) depends entirely on the particular specification of the model.

For example, Joel Schwartz reported a statistically significant association between the level of coarse particulate concentration the day before death (a one-day lag) and mortality rates in Birmingham, Alabama, in statistical models where the explanatory variables included only this measure of coarse particulates and temperature.[43] Looking at updated Birmingham data, Richard L. Smith and coauthors replicated this result, but found that when they included measures of coarse particulates for the day of exposure plus the four previous days, coarse particulate exposure did not have a statistically significant effect on mortality.[44] Moreover, when they included a variable for humidity in the Schwartz model, the estimated effect of coarse particulate exposure was much smaller than when only temperature was included as an explanatory environmental variable.

The model selection bias is ubiquitous in all studies of the health effects of air pollution, regardless of the pollutant under consideration. This bias is related to the p-hacking problem discussed earlier. S. Stanley Young recently summarized the shaky statistical basis of such studies:

> Environmental epidemiology typically has few, if any, analysis requirements. In an environmental observational (EO) study, the researcher can modify the analysis as the data is examined. Multiple outcomes can be examined, multiple variables (air components) can be used as predictors. The analysis can be adjusted by putting multiple covariates into and out of the model. It is thought that effects can be due to events on prior days so different lags can be examined. $PM_{2.5}$ yesterday or the day before can cause deaths today. Seldom, if ever, is there a written, statistical protocol prior to examination of the data. With these factors (outcomes, predictors, covariates, lags), there is no standard analysis strategy. The strategy can be try-this-and-try-that.[45]

Two other studies vividly illustrate how model specification bias persists in some of the key studies underlying the EPA's 2009 fine PM assessment.[46] Over the past two decades, at a national level in the United States, both fine PM levels and cardiovascular mortality rates among people 65 years of age and older have fallen (see Figure 11.6).[47]

Tony Cox and Douglas Popken found that changes in county population, age, and external-cause mortality all were statistically significant in explaining changes in both all-cause and cardiovascular mortality, but when these variables were included, adding changes in fine PM did not add to the model's explanatory power. In contrast, when the dependent variable was raw mortality data, and year (time) and fine PM the predictors, fine PM had a small but statistically significant role. If the confounding effects of the demographic variables are so strong as to destroy this relationship, any such model for fine PM effects has little utility.

FIGURE 11.6

Trends in PM₂.₅ and mortality rates

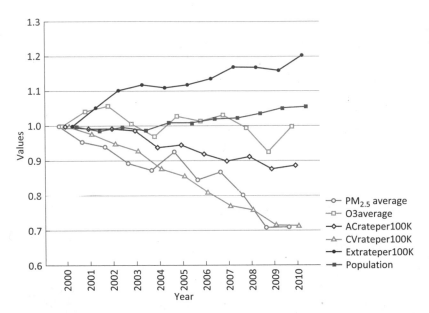

Source: Louis Anthony Cox and Douglas A. Popken, "Has Reducing Fine Particulate Matter and Ozone Reduced Mortality Rates in the United States?" *Annals of Epidemiology* 25 (2015): 166.

Note: CV is death due to heart/circulatory disease; AC is all-cause mortality; Ext is death due to external causes such as accidents. Average ambient Ozone (O3) is also included.

Another illustration of model selection bias is depicted in Figure 11.7. The data are obviously so noisy as to be useless. In the far lower-left corner of the graph, one can see that city number 46, Topeka, is a clear outlier. Removal of that single outlier data point destroys the statistical significance of the purported linear relationship.[48]

It is not possible for "national" studies to be truly nationwide. Examining a subset of the data used by Pope and coauthors, S. Stanley Young and Jessie Q. Xia found that the data showed an increase in life expectancy with a decrease in fine PM only in the eastern United States, not in the western United States.[49] They also found that in both the eastern and western United States, an increase in income over the study period was the most important variable for predicting decreased

FIGURE 11.7

Estimated increase in life expectancy from a reduction in PM$_{2.5}$ of 10 µg/m³ for 51 U.S. metropolitan areas

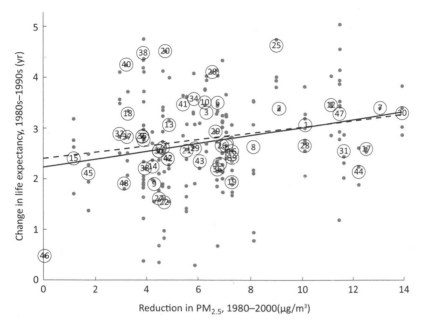

Source: C. A. Pope III, M. Ezzati, and D. W. Dockery, "Fine-Particulate Air Pollution and Life Expectancy in the United States," *New England Journal of Medicine* 360 (2009): 376–386.

Note: There were 51 measuring stations, which Pope et al. replicated as necessary and then assigned to 211 metropolitan areas.

mortality, not fine PM. If regulators wish to reduce the mortality rate, then at the current margin they might do well to act in ways that increase Americans' incomes.

In another testament to the weakness of any relationship between fine PM and mortality (if there even is one), Young and coauthors subjected the Pope data to a cluster analysis to isolate common responses. Of the ten resultant clusters, only one displayed a significant relationship between fine PM and mortality.[50]

Finally, Cox and coauthors asked whether fine PM has *any* power to help explain or predict mortality rates *after* the analysis takes into account all the other variables that do predict mortality. Of the 101 cities they studied, after conditioning on month of the year—the strongest predictor of mortality rates—in only six cities was fine PM a significant predictor of mortality. As they summarize this, "*It does not appear that current or lagged PM$_{2.5}$ values have statistically significant positive associations with all-cause or cause-specific mortality rates, after conditioning on month and temperature-related variables*" (emphasis in original).[51]

It is patently obvious that the fine PM mortality studies do not make a case for nationwide regulation.

The Neglected Possibility That Fine PM Exposure Could Decrease Risk

In a large number of instances the error bars are so wide that many cities can display a negative mortality effect (see Figures 11.3 and 11.4). The discussion here strongly suggests that when the full set of variables explaining variation in mortality are included—in particular the season of the year, correctly measured temperature, individual lifestyle, and neighborhood socioeconomic variables—then fine PM may have no independent power in explaining mortality.

The researchers whose work the EPA relies on clearly believe that any estimated negative relationship between fine PM levels and mortality must be spurious. As Borek Puza and coauthors put it, "A problem of interpretation can arise if a CI [confidence interval] containing negative values leads people to believe that air pollution might be beneficial to health."[52]

However, as noted in Chapter 7 by Edward Calabrese in this volume,[53] a large number of studies have found that for ionizing radiation,[54] chemicals, and pollutants, very low levels of exposure can indeed reduce risk. Those studies find that the expected positive relationship—increases in concentration of the chemical or pollutant that raise the probability of an adverse health outcome—only arises for higher levels of concentration. This U-shaped relationship is called hormesis.

FIGURE 11.8

Annual mortality rates from cardiovascular disease (A) and pneumonia/
influenza (B) and fine PM levels in 10 U.S. cities

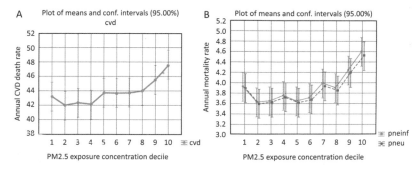

Source: Reproduced from Louis Anthony Cox Jr., "Hormesis for Fine Particulate Matter (PM 2.5),"
Dose Response 10 (2012): Figure 1.

Hormesis may describe the relationship between fine PM and mortality. This might seem counterintuitive, for as EPA researchers continually emphasize, fine PM particles are able to penetrate into the lungs. However, as Cox notes, very low levels of fine PM exposure are associated with an increased production of antioxidants. Only at higher concentration levels may the balance tip in favor of harmful oxidants that increase the risk of a variety of lung diseases.[55]

When Cox looked at data on daily mortality and fine PM concentration levels in the ten largest U.S. metropolitan areas over the period 1987–2000, he found a relationship that is not inconsistent with hormesis for cardiovascular death and death from pneumonia/influenza.[56] These relationships are depicted in Figure 11.8.

Much more work needs to be done to investigate the possibility of a hormetic relationship between fine PM concentration levels and mortality from various causes. A simple test could involve agriculture, where soils release large amounts of both fine and coarse particulate matter when disturbed and deflated.[57] Farmers are exposed to enormous concentrations of fine PM when tilling dry soil, especially when it is windy. A study of farmers, a human population that is routinely exposed to high and varying levels of both coarse and fine particulates, would surely yield valuable information. The EPA has yet to fund any such research.

The EPA-Funded Risk Production Complex

In light of the foregoing discussion, it may seem surprising, and perhaps even shocking, that since 1997 the EPA has continued to assert in its scientific assessment that fine PM exposure causes increased mortality risk. Indeed, in its 2009 fine PM assessment, the EPA mentions *none* of the studies that had been published before 2009 and which therefore could well have been discussed in the 2009 assessment. In addition, Suresh Moolgavkar published a comprehensive critique of the work the EPA had relied on until 2005; this is also not mentioned in the EPA's 2009 assessment.[58]

Advocacy, Not Assessment

This is not because the EPA was unaware of the critical work. As Moolgavkar explains, a meta-analysis by Gary Koop and Lise Tole clearly called into question the fine PM–mortality relationship,[59] and this "was brought to the attention of the Agency during the last review . . . in early 2004, both by me and, more importantly, by a member of the CASAC [Clean Air Science Advisory Committee]. However, the current version . . . continues to ignore the paper."[60]

According to Moolgavkar, a 2004 article by Koop and Tole, as well as other published articles using methodologies to test whether statistical results depend on a particular, arbitrary model specifications (in particular Bayesian Model Averaging) were brought to the EPA's attention as early as 2004. However, these were only "perfunctorily discussed by the EPA in its 2004 draft of the Criteria Document." He went on:

> In my opinion, [Bayesian Model Averaging] deserves a lot more attention because of its implications for the conclusions drawn from epidemiological data, particularly when estimated risks are very small as is the case for air pollution.[61]

The 2004 Koop and Tole paper was challenged thoroughly and systematically by Duncan C. Thomas and coauthors, who were a "veritable Who's Who of air pollution epidemiologists and statisticians." Their paper was circulated to the EPA, long before its publication in the refereed literature in 2004, and before the agency's 2005 fine PM assessment was produced.[62]

Thomas and coauthors found unconvincing Koop and Tole's failure to find that $PM_{2.5}$ had a statistically significant positive effect on mortality. In particular, they stressed that Koop and Tole might have found that $PM_{2.5}$ had no statistically significant impact because people stay indoors and so suffer fewer adverse health effects from heavy air pollution. If this were true, then it would be true of all epidemiological studies of fine PM, not just those that attempt to evaluate multiple models rather than only one or two. Thomas and coauthors also noted that choosing a particular statistical model requires substantive knowledge about how the environment affects human health. However, in the opinion of one very experienced epidemiologist, "I wish we had 'substantive knowledge about environmental health processes.' The fact is we do not. There is no biological basis for choice of lag structure, nor for appropriate models for control of weather and temporal trends."[63]

In the end, Thomas and coauthors stuck with the view that increases in $PM_{2.5}$ increase mortality for the simple reason that the "majority of the evidence" from multicity studies shows that "exposure to ambient air pollution" is associated with significant elevations in mortality. As Pope and coauthors put this justification more recently, "Associations between long-term exposure to fine particulate air pollution and mortality have been observed . . . more recently, in cohort-based studies . . . [and] all support the view that relatively prompt and sustained health benefits are derived from improved air quality."[64]

In other words, according to these researchers, because they had published so many papers showing a statistically significant positive relationship between fine PM and mortality, that relationship cannot be disconfirmed by any subsequent evidence, even when obvious systematic and statistical flaws are brought to light. This is not the way the scientific method works, and its proclamation is stark testament to the current sickness of modern regulatory science.

The CASAC is supposed to ensure that the EPA Assessment Reports reflect what is learned from the scientific method. Under the CAA, the CASAC is to be an "independent scientific review committee." By 2017, the CASAC had long been the opposite of an independent review group. The dozens of studies reporting a positive association between fine PM and mortality have been funded by the EPA and reviewed by an EPA CASAC, whose members are precisely the same researchers whom the EPA has funded. Far from being independent, the majority of CASAC members are heavily dependent on EPA funding. To expect the EPA or its CASAC to objectively and critically assess studies that their members have produced is to expect a superhuman degree of detachment.

In the case of fine PM, the research that they produced is the primary basis for finding a causal relationship in the fine PM Assessment. The EPA pays these scientists both to produce research and then to review their own research. Such review is taken as providing support for regulatory action in setting revised air quality standards.

As this inherently conflicted structure seems to have gone relatively unnoticed, it is worthwhile to review its history and organization. In 1998, the National Research Council (NRC) convened a Committee on Research Priorities for Airborne Particulate Matter. That same year, the committee issued the first of what became five reports. That initial report, "Immediate Priorities and a Long-Range Research Portfolio," recommended that the EPA create five "airborne particulate matter research centers" ("PM Centers") to "address research needs in the areas of exposure, dosimetry and extrapolation modeling, toxicology, and epidemiology."[65]

The EPA followed the recommendations of this report, creating five fine particulate centers in 1998: the Northwest Research Center for Particulate Air Pollution and Health at the University of Washington, which received an EPA grant of $8,288,977;[66] the EPA NYU PM Center: Health Risks of PM Components, which received an EPA grant of $8,076,438;[67] the Rochester PM Center at the University of Rochester, which received an EPA grant of $8,302,447;[68] the Southern California

Particle Center and Supersite at UCLA, which received a grant of $8,715,583;[69] and the Harvard Center for Ambient Particle Effects at Harvard University, which received a grant of $7,747,040.[70] The research grants received by these centers were extended to 2006. Thus, over the period 1999–2006, the EPA invested about $40 million in five centers to investigate the health effects of fine PM. Perhaps not coincidentally, all of these (except apparently the UCLA center) were located at the institutional homes of members of the 1998 NRC Committee.[71]

When the original particle center grants from the EPA officially terminated in 2005, the agency funded the continuation of three of the particle centers, at Harvard (for $7,999,609),[72] Rochester (for $8,000,000),[73] and UCLA (for $7,999,994)[74] for the period 2005–2010. All three grants were extended to 2012. In addition, two new centers were established for the period 2005–2010: the Johns Hopkins Particulate Matter Research Center received $7,993,275;[75] and the San Joaquin Valley Aerosol Health Effects Research Center (SAHERC) at the University of California, Davis, received $7,999,767.[76] Total center funding was around $80 million.

These grants were only part of the EPA's investment in research into the health effects of fine PM. In addition to funding these academic centers, the EPA has consistently and massively funded the Health Effects Institute (HEI). Billed as a public-private partnership, the HEI funded many of the fine PM–mortality studies that the EPA relies on for causation. Over the period from 2000 to 2019, the HEI has received at least $87 million in grants from the EPA.[77]

Between the fine PM centers and the Health Effects Institute, the EPA has invested almost $180 million in research exploring the health effects of fine PM. Even this is an understatement, however, for it also invested about $30 million in a 2004–2014 multi-university, multi-ethnic study of atherosclerosis, cardiovascular disease, and long-term PM exposure called the MESA Air Study.[78] Including this study, the EPA has provided over $210 million since 1999 to fund research investigating the health effects of exposure to PM.

By performing the research, summarizing what they deem relevant, and then recommending expensive policies, these centers are judge, jury, and executioner in one person.

To see how the centers are producing the research that supports the EPA's finding of a causal connection between fine PM and mortality, consider the key short- and long-term studies that the EPA relied on in its 2009 fine PM ISA. It gives enormous weight to national, short-term, multi-city studies by Franklin and coauthors, and by Antonella Zanobetti and Joel Schwartz.[79] All of these studies were funded by the EPA; two were funded by the Harvard Particulate Center, and one by the National Institute of Environmental Health Sciences plus an EPA Star Grant. As for the authors, since 1999, Schwartz has been a principal of the Harvard PM center and its follow-up centers; Schwartz and Zanobetti (who is also a researcher at the Harvard center) are authors of numerous articles supported by that EPA-funded center. All other coauthors were graduate students or researchers at the Harvard PM center.

As for the long-term studies the EPA relies on, by far the most important are various analytical permutations (analyses, reanalyses, re-reanalyses) of the long-term American Cancer Society cohort and the Harvard Six Cities study. All of these studies were funded by the agency.[80] The massive 2009 study by Krewski and coauthors, a reanalysis and extended follow-up of the ACS cohort, was funded by the HEI, which, as noted, has itself received close to $100 million in funding from the EPA since 2000. As for the authors, although principal investigator Pope's home is Brigham Young University, from 2000 to 2004 he was chair of the scientific advisory committee of Harvard's PM center and a member of New York University's PM center scientific advisory committee, while continuing to serve on the scientific advisory committees of Harvard and NYU and on the same committee at Johns Hopkins for the period 2006–2010; for the entire period 2003–2011 he was on HEI's international scientific advisory committee.

Table 11.1 in the Appendix names the members of the 2005 CASAC fine PM review panels along with their institutional affiliations and lower bounds on EPA funding for their affiliated institutions. It begins with fine PM review panel members who were also members of CASAC. Table 11.2, also in the Appendix, provides the same information for the EPA's 2010 fine PM review panel. Note that in both tables the funding amounts include both the institutional funding and any individual grants I have been able to identify.

This massive integrated inbreeding is a recent development. If we look first at the 2005 CASAC fine PM review committee, we see that of the nine academic members who were not members of CASAC itself but on CASAC's fine PM review committee because of their expertise in fine PM research, *almost all*—eight out of nine—were affiliated with the EPA-created fine PM centers, and *every EPA fine PM center director was on the panel*. On the other hand, the EPA's 2005 fine PM panel did include seven nonacademics, at least five of whom had no history of receiving EPA funding.

This relative balance on the 2005 fine PM review panel still produced a biased and incomplete result. That panel documented the new studies (done after the EPA's assessment report) brought to the panel's attention by members. The new studies brought to the attention of the full panel by EPA-funded members all cut in only one direction—that the causal connection between fine PM and mortality was even stronger than previous work suggested.[81] Conversely, panel members without an EPA affiliation, such as George Wolff of General Motors, suggested that the EPA's scientific assessment should include new studies questioning the fine PM–mortality causal link. Wolff brought Moolgavkar's 2005 article to the committee's attention, aptly summarizing that paper as concluding that "a particle mass standard is not defensible on the basis of a causal connection between ambient particle mass and adverse effects on human health."[82]

As can be seen by comparing Tables 11.1 and 11.2, the changes between 2005 and 2010 essentially *eliminated from EPA's fine PM review panel any scientist who was not EPA-funded and/or affiliated with an EPA center*. The mission statements of all of these centers make it abundantly clear that their objective was not to investigate *whether* fine PM had health effects, but to produce studies documenting the size and nature of those effects.[83]

By the time of the 2010 fine PM review, the EPA's CASAC fine PM review panel included only EPA-funded academics and contained precisely zero members, regardless of affiliation, who had either the knowledge or the interest to question underlying EPA-funded research supposedly under review. Instead, the panel was dominated by academic researchers whose EPA-funded work was precisely the work being reviewed. It is hard to see how such a review panel could be described as independent. The EPA funds the production of centers that produce studies showing causation and then places the academic creators of such research on the review boards that are supposed to objectively critique the studies.

This is a situation remarkably similar to the relationship between the U.S. Global Change Research Program (USGCRP) and the mandated quadrennial National Assessments of the effects of climate change on the United States. The USGCRP is composed precisely of the agencies and entities that are funded to study climate change, and the agencies surely know if they find or state that the effects are less than catastrophic, the funding will vanish. Similar results accrue if the truth is told about fine PM—that the data are far too noisy and the analyses far too flawed to support strict reductions. But given the incentives of the academy, monopolistic government funding for certain fields of science fraught with regulatory implications turns science on its head. Rather than ensuring that regulation is based on science produced by the scientific process, such funding generates "scientific" studies that establish a persistent justification for regulation regardless of whether their conclusions would survive further scientific scrutiny.

TABLE 11.1

2005 EPA fine PM review panel membership and institutional affiliations

Name	Institutional affiliations	EPA funding relationships
CASAC members		
Rogene Henderson (Chair of CASAC and Chair of fine review panel)	Lovelace Respiratory Institute Health Effects Institute (HEI) (Chair of HEI research committee, 1998–2005)	Lovelace $4 million HEI $18.75 million grant
Ellis Cowling	North Carolina State University	Developed EPA funded National Atmospheric Deposition Program for acid rain monitoring
James Crapo	University of Colorado, Denver Medical Center	
Frank Speizer	Harvard PM Center	$16 million
Fine PM panel members appointed due to expertise in fine PM epidemiology or toxicology		
Philip Hopke	Clarkson University, Rochester University/ EPA PM Center	Clarkson $844,000 Rochester PM Center $16 million
Jane Koenig	University of Washington/EPA Center for Particulate Air Pollution and Health (Director) and HEI	$27 million
Petros Koutrakis	Harvard PM Center	$24 million
Morton Lippman	NYU EPA PM Center (Director)	$28 million

(*continued*)

TABLE 11.1

(continued)

Name	Institutional affiliations	EPA funding relationships
Gunter Obersdorster	Rochester EPA PM Center (Director)	$18 million
Jonathan Samet	Johns Hopkins PM Research Center (Director) and HEI	$27 million
Sverre Vidal	University of Washington/EPA Center for Clean Air Research (Director)	$8 million
Ron White	Johns Hopkins PM Research Center	
Warren White	UC Davis Crocker National Laboratory	Since 1985, Crocker Laboratory partnered with the EPA to develop nationwide Class I air quality monitoring network
Nonacademic fine PM review panel members		
Allan Legge	Biosphere Solutions, Inc.	
Joe Mauderly	National Respiratory Center	EPA funded center
Roger McClellan	Consultant	
Frederick Miller	Consultant	
Robert Rowe	Stratus Consulting	Long history of EPA consulting contracts
George Wolff	Chief Scientist, General Motors Corp.	
Barbara Zielinska	Desert Research Institute	

TABLE 11.2

2010 EPA fine PM review panel membership and institutional affiliations

Name	Institutional affiliations	Lower bound on EPA funding relationships
CASAC members		
Armistead Russell (CASAC Chair)	Georgia Tech, Georgia Tech/Emory "Southeastern Center for Air Pollution and Epidemiology"	$11 million
Christopher Frey	North Carolina State	$3 million
Richard Poirot	Vermont Department of Environmental Conservation	
Kathleen Weathers	Cary Institute for Ecosystem Research	$3 million
Fine PM panel members appointed due to expertise in fine PM epidemiology or toxicology		
Lowell Ashbaugh	UC Davis Crocker National Laboratory	See Table 11.1
Ed Avol	USC, Southern California Particle Center and Supersite	$44 million
Joe Brain	Harvard PM Center	$24 million
Rogene Henderson	See Table 11.1	See Table 11.1
Philip Hopke	See Table 11.1	See Table 11.1
Morton Lippman	See Table 11.1	See Table 11.1
Robert Phalen	UC Irvine, Southern California Particle Center and Supersite	$8 million
Kent Pinkerton	UC Davis, San Joaquin PM Center	$10 million
Frank Speizer	See Table 11.1	See Table 11.1
Helen Suh	Harvard PM Center	$24 million
Sverre Vidal	See Table 11.1	See Table 11.1

NOTES

Introduction

Note that primary references cited in this chapter are those not cited in later chapters.

1. M. Kowalski et al., "Helicobacter pylori Infection in Coronary Artery Disease," *Journal of Physiology and Pharmacology* 57, Supplement 3 (2006): 101–111.

2. Select Committee on Nutrition and Human Needs, *Dietary Goals for the United States*, U.S. Senate (Washington: Government Printing Office, 1977), Internet Archive, https://archive.org/details/CAT10527234.

3. Merely replacing the words "the reduction of salt intake for" with "the effects of global warming on" creates the stock Gore language on climate change.

4. Kevin Dayaratna, Ross McKitrick, and David Kreutzer, "Empirically-Constrained Climate Sensitivity and the Social Cost of Carbon," April 5, 2016, https://doi.org/10.2139/ssrn.2759505.

5. NEEM Community Members, "Eemian Interglacial Reconstructed from a Greenland Folded Ice Core," *Nature* 493 (January 24, 2013): 489–494, https://www.nature.com/articles/nature11789.

6. F. Hourdin et al., "The Art and Science of Climate Model Tuning," *Bulletin of the American Meteorological Society* 98 (March 31, 2017), https://doi.org/10.1175/BAMS-D-15-00135.1.

7. Pruitt resigned as EPA administrator in the Trump administration in July 2018.

8. Charles A. Murray, *Losing Ground: American Social Policy, 1950–1980* (New York: Basic Books, 1984).

Chapter 1

1. Francis Bacon, *Cogitata et Visa* (1607) quoted in William Leiss, *The Domination of Nature* (Montreal: McGill Queen's University Press, 1994), p. 50.

2. Francis Bacon, *Cogitata et Visa* (1607) quoted in Sophie Weeks, "The Role of Mechanics in Francis Bacon's Great Instauration," in Claus Zittel et al., *Philosophies of Technology: Francis Bacon and His Contemporaries* (Leiden and Boston: Brill, 2008), p. 182.

3. Martin Luther (June 4, 1539) quoted by Carlos Eire in *Reformations: The Early Modern World, 1450–1650* (New Haven: Yale University Press, 2016), p. 678.

4. Herman J. Selderhuis, *The Calvin Handbook* (Grand Rapids, MI: William B. Eerdmans, 2009), p. 452.

5. Jonathan Swift, *The Works of Jonathan Swift, D.D.* (1726; repr. Edinburgh: Archibald Constable, 1814), pp. 229–230.

6. Francis Bacon, *Advancement of Learning*, in *The Works of Francis Bacon*, vol. 2, p. 1, https://en.wikisource.org/wiki/Index:The_Works_of_Francis _Bacon_(1884)_Volume_2.djvu/.

7. Today people claim science to be a public good that, in the jargon, is nonrivalrous and nonexcludable: 400 years ago Bacon intended the same meanings when he described it as a "universality."

8. Francis Bacon, *De Augmentis Scientiarum* (1623), in *The Works of Francis Bacon*, vol. 4, ed. James Spelling, Robert Ellis, and Douglas Heath (London: Longman, 1858), p. 408.

9. Walter Eltis, "The Contrasting Theories of Industrialization of Francois Quesnay and Adam Smith," *Oxford Economic Papers* 40 (1988): 269–288.

10. Adam Smith, *(1772–73) Lectures on Jurisprudence*, ed. R. L. Meek, D. D. Raphael and P. G. Stein, vol. 5 of the *Glasgow Edition of the Works and Correspondence of Adam Smith* (Indianapolis: Liberty Fund, 1982), p. 336.

11. Quoted in Nathan Rosenberg, "Karl Marx on the Economic Role of Science," *Journal of Political Economy* 82 (1974): 713–728.

12. Joseph Schumpeter, *Capitalism, Socialism and Democracy* (1942; repr. London: Routledge, 1994), p. 83.

13. Bacon, *Advancement of Learning*, vol. 1, p. 70.

14. Adam Smith, *Wealth of Nations*, ed. J. M. McCullock, Adam Black, Charles Black, and William Tait (1776; repr. Edinburgh: n.p., 1838), p. 347.

15. Adam Smith, *Wealth of Nations* (1776; repr. Edinburgh: Thomas Nelson and Peter Brown, 1827), p. 6.

16. Edwin Mansfield, "Academic Research and Industrial Innovation," *Research Policy* 20 (1991): 1–12.

17. David Hume, *Political Essays* (1777) quoted in Knud Haakonsen, *Hume: Political Essays* (Cambridge: Cambridge University Press, 1994), p. 30.

18. The NAS had its informal or invisible precursor in the Lazzaroni, an informal group of scientists who lobbied for a science academy and who were led by Alexander Bache, the engineer who was the superintendent of the U.S. Coast Survey and the great-grandson of Benjamin Franklin. Other academy precursors in the United States included the American Philosophical Society (founded in Philadelphia in 1743 by Benjamin Franklin, among others), the American Academy of Arts and Sciences (founded in Cambridge, MA, in 1780), and the Franklin Institute (founded in Philadelphia in 1824), but they leaned British in that they did not aspire to closeness to government: the NAS leaned continental European in seeking close alignment with the federal government.

19. Eric Jones, *The European Miracle*, 3rd ed. (Cambridge: Cambridge University Press, 2003), p. 218.

20. James Buchanan, "Veto Message Regarding Land-Grant Colleges," Miller Center, University of Virginia, Charlottesville, February 24, 1859, https://miller center.org/the-presidency/presidential-speeches/february-24-1859-veto-message -regarding-land-grant-colleges.

21. Terry Reynolds, "The Education of Engineers in America before the Morrill Act of 1862," *History of Education Quarterly* 32 (1992): 459–482.

22. Terence Kealey, *The Economic Laws of Scientific Research* (London: Macmillan, 1996), p. 151.

23. University of Groningen, Maddison Project Database, version 2018; 1990$ benchmark, https://www.rug.nl/ggdc/historicaldevelopment/maddison/releases /maddison-project-database-2018; Jutta Bolt et al., "Rebasing 'Maddison': New Income Comparisons and the Shape of Long-run Economic Development," Maddison Project Working Paper no. 10, 2018, www.ggdc.net/maddison.

24. Maddison Project Database, version 2018; Bolt et al., "Rebasing 'Maddison.'"

25. Maddison Project Database; Bolt et al.

26. OECD, *The Sources of Economic Growth in OECD Countries* (Paris: OECD, 2003).

27. Walter Park, "International R&D Spillovers and OECD Economic Growth," *Economic Inquiry* 33 (1995): 571–591.

28. Leo Sveikauskas, *R&D and Productivity Growth: A Review of the Literature* (Washington: Bureau of Labor Statistics, 2007), www.bls.gov/osmr/pdf /ec070070.pdf.

29. Daniele Archiburgi and Andrew Filippetti, "The Retreat of Public Research and Its Adverse Consequences on Innovation," *Technological Forecasting and Social Change* 127 (2018): 97–111.

30. Edwin Mansfield, Mark Schwartz, and Samuel Wagner, "Imitation Costs and Patents: An Empirical Study," *Economic Journal* 91 (1981): 907–918.

31. Richard Levin et al., "Appropriating the Returns from Industrial Research and Development," *Brookings Papers on Economic Activity* 18 (1987): 783–820.

32. Nathan Rosenberg, "Why Do Firms Do Basic Research (with Their Own Money)?" *Research Policy* 19 (1980): 165–174; Wesley M. Cohen and Daniel Levinthal, "Innovation and Learning: The Two Faces of R&D," *Economic Journal* 99 (1989): 569–596.

33. Christopher Freeman and Luc Soete, *The Economics of Industrial Innovation* (London: Pinter, 1997); Diana Hicks and J. S. Katz, *The Changing Shape of British Industrial Research*, STEEP Special Report no. 6, University of Sussex, UK.

34. Paula Stephan, "The Economics of Science," *Journal of Economic Literature* 34 (1996): 1199–1235.

35. Zvi Griliches, "Productivity, R&D, and Basic Research at the Firm Level in the 1970s," *American Economic Review* 76 (1986): 141–154; Edwin Mansfield, "Basic Research and Productivity Increase in Manufacturing," *American Economic Review* 70 (1980): 863–873.

36. George Stigler, "The Economics of Information," *Journal of Political Economy* 69 (1961): 213–225.

37. Copernicus (1543), *De revolutionibus*, quoted by Steven Shapin in *The Scientific Revolution* (Chicago: University of Chicago Press, 1996), p. 122.

38. Denis Diderot, "Arts" in the *Encyclopedie*, quoted by Joel Mokyr in *Gifts of Athena: Historical Origins of the Knowledge Economy* (Princeton: Princeton University Press), p. 28.

39. Michael Polanyi, *Personal Knowledge: Towards a Post-Critical Philosophy* (Chicago: University of Chicago Press, 1958), pp. 4, 55, 95.

40. Levin et al., "Appropriating the Returns from Industrial Research and Development."

41. A. J. Lotka, "The Frequency Distribution of Scientific Productivity," *Journal of the Washington Academy of Sciences* 16 (1926): 317–323.

42. Craig Venter, *A Life Decoded* (New York: Viking, 2007), p. 142.

43. H. M. Collins, "The TEA Set: Tacit Knowledge and Scientific Networks," *Science Studies* 4 (1974): 165–186.

44. Collins, "The TEA Set."

45. Steven Shapin and Simon Schaffer, *Leviathan and the Air-Pump: Hobbes, Boyle and the Experimental Life* (Princeton: Princeton University Press, 1985).

46. Terence Kealey and Martin Ricketts, "Modelling Science as a Contribution Good," *Research Policy* 43 (2014): 1014–1024.

47. Eric von Hippel, "The Economics of Product Development by Users," *Management Science* 44 (1998): 629–644.

48. Thomas J. Allen, Diane B. Hyman, and David L. Pinckney, "Transferring Technology to the Small Manufacturing Firm," *Research Policy* 12 (1983): 202.

49. Albert N. Link and Laura L. Bauer, *Cooperative Research in U.S. Manufacturing* (New York: Lexington Books, 1989); John Scott, "Environmental Research Joint Ventures among Manufacturers," *Review of Industrial Organization* 11 (1996): 655–679.

50. Scientists proselytize a false consensus because, of course, they hope to benefit from it; they would rather receive vast funds from government for which they account only to each other than be subject to the budgets and direction of companies. That is, scientists are not disinterested players, and a version of public choice theory applies to them.

51. Modern scholarship confirms that wealth and power, today, still emerge from technical change in the widest sense. Robert Solow, "Technical Change and the Aggregate Production Function," *Review of Economics and Statistics* 39 (1957): 312–320.

52. Francis Bacon, *Novum Organum* (1620) cited in Adam Lucas, "Narratives of Technological Revolution," in *Handbook of Medieval Studies*, ed. Albrecht Classen (Berlin and New York: de Gruyter, 2010), p. 968.

53. Peter Russell, *Prince Henry "the Navigator": A Life* (New Haven: Yale University Press, 2000).

54. The letter to Roosevelt was actually written by Leo Szilárd, the physicist who fled Germany a year earlier. The letter is archived in the Franklin Delano Roosevelt Presidential Library and is reproduced by *Wikipedia* in its entry "Einstein–Szilárd letter."

55. The project was first based in Manhattan, hence the name.

56. Given the state of his health, Roosevelt may have directed Bush to write the letter for his signature.

57. Quoted in Vannevar Bush, *Science: The Endless Frontier* (Washington: Government Printing Office, 1945), pp. 3–4.

58. Bush, *Science*, p. 4.

59. Sometimes satirized as "Science: The Endless Budget."

60. Cited in *Science Policy Background Report No. 1* (Washington: Government Printing Office, 1986), p. 30.

61. Pax Britannica ended at 9:00 a.m., February 24, 1947, when Lord Inverchapel, the British ambassador in Washington, handed Secretary of State George Marshall two aide-mémoires asking the United States to assume responsibility for Greece and Turkey: the British were scuttling. Six months later the British left India/Pakistan, and a year later they left Palestine/Israel.

62. Kealey, *Economic Laws of Scientific Research*, pp. 151, 162.

63. Brian Nosek et al., "Estimating the Reproducibility of Psychological Science, *Science* 349: aac4716.

64. Daniel Gilbert et al., "Comment on 'Estimating the Reproducibility of Psychological Science,'" *Science* 351 (2016): 1037, https://doi.org/10.1126/science.aad7243.

65. Paul E. Smaldino and Richard McElreath, "The Natural Selection of Bad Science," *Royal Society Open Science*, https://doi.org/10.1098/rsos.160384.

66. Daniele Fanelli, "Negative Results Are Disappearing from Most Disciplines and Countries, *Scientometrics* 90 (2012): 891–904.

67. John P. A. Ioannidis, "Why Most Published Research Results Are False, *PLoS Medicine* 2 (2005): e124, https://doi.org/10.1371/journal.pmed.0020124.

68. D. Fanelli and John P. A. Ioannidis, "U.S. Studies May Overestimate Effect Sizes in Softer Research," *Proceedings of the National Academy of Sciences* 110 (2013): 15031–15036.

69. Silas B. Nissen et al., "Publication Bias and the Canonization of False Facts," *eLife* (2016): e21451, https://doi.org/10.7554/eLife.21451.

70. Hans-Hermann Dubben, "Systematic Review of Publication Bias in Studies on Publication Bias," BMJ 331 (2005): 433.

71. H. L. Head et al., "The Extent and Consequences of p-hacking in Science," *PLoS Biology* 13 (2015), https://doi.org/10.1371/journal.pbio.1002016.

72. G. D. Smith and S. Ebrahim, "Data Dredging, Bias or Confounding," *British Medical Journal* 325 (2002): 1437–1438.

73. N. L. Kerr, "HARKing: Hypothesizing after the Results Are Known," *Personality and Social Psychology Review* 2 (1998): 196–217.

74. Daniele Fanelli, Rodrigo Costas, and John P. A. Ioannidis, "Meta-Assessment of Bias in Science," *Proceedings of the National Academy of Sciences* 114 (2017): 3714–3719.

75. Daniel Sarewitz, "Saving Science," *The New Atlantis* 49 (Spring/Summer 2016): 4–40.

76. Dwight D. Eisenhower, Transcript of Dwight D. Eisenhower's Farewell Address, 1961, https://www.ourdocuments.gov/doc.php?flash=false&doc=90&page=transcript.

77. Fred Stone, quoted in Paula Stephan, "The Endless Frontier: Reaping What Bush Sowed?" in *The Changing Frontier: Rethinking Science and Innovation Policy*, ed. Adam Jaffe and Benjamin Jones (Chicago: University of Chicago Press, 2015), p. 326.

78. R. G. Steen, A. Casadevall, and F. C. Fang, "Why Has the Number of Scientific Retractions Increased?" *PLoS ONE* 8 (2013): e68397.

79. Lotka, "The Frequency Distribution of Scientific Productivity."

80. Steve Kolowich, "The Water Next Time: Professor Who Helped Expose Crisis in Flint Says Public Science Is Broken," *Chronicle of Higher Education*, February 2, 2016, http://chronicle.com/article/The-Water-Next-Time-Professor/235136.

81. Kolowich, "The Water Next Time."

82. Kolowich.

83. The vast majority of universities make an "up or out" decision during a faculty member's sixth year. Those who do not merit tenure (appointment without term) are not renewed.

84. An example: one of us (Michaels) published an op-ed on climate change, and it found itself on the (heavily funded via climate change) department chairman's desk, with "What's he going to do to our NSF funding? We've got to stop him" scrawled over it.

85. Thomas Jefferson, letter from Thomas Jefferson to John Adams, September 10, 1819, https://founders.archives.gov/documents/Jefferson/98-01-02-0953.

86. Earlier, as governor of the state of Virginia, Jefferson had used his influence to abolish William and Mary's school of divinity and to abort its endowed program of reaching out to Native Americans (then called Indians); William and Mary, "Jefferson's Attempts at Change," http://www.wm.edu/about/history/tjcollege/tjattemptsatchange/index.php. As he revealed in the Declaration of Independence, Jefferson hated Native Americans.

Chapter 2

1. Select Committee on Nutrition and Human Needs, *Dietary Goals for the United States*, U.S. Senate (Washington: Government Printing Office, 1977), p. 1, Internet Archive, https://archive.org/details/CAT10527234.

2. Senate Select Committee, *Dietary Goals*, p. v.

3. Senate Select Committee, p. 12.

4. Department of Health, Education and Welfare, *Vital Statistics of the United States, 1968: Vol. 2: Mortality, Part A* (Rockville, MD: Public Health Service, 1972), pp. 1–6, www.cdc.gov/nchs/data/vsus/mort68_2a.pdf.

5. Leon Michaels, "Aetiology of Coronary Artery Disease: An Historical Approach," *Heart* 28 (March 1966): 258–264, https://doi.org/10.1136/hrt.28.2.258.

6. Marc LaLonde, *A New Perspective on the Health of Canadians* (Ottawa, Ontario: Minister of Supply and Services Canada), Public Health Agency of Canada website, http://www.phac-aspc.gc.ca/ph-sp/pdf/perspect-eng.pdf.

7. Ancel Keys, "Atherosclerosis: A Problem in Newer Public Health," *Journal of the Mount Sinai Hospital* 2 (July and August 1953): 118–139.

8. Nina Teicholz, *Big Fat Surprise* (London: Scribe, 2014), p. 36.

9. Study Group on Atherosclerosis and Ischaemic Heart Disease, *Proceedings* 117 (Geneva: World Health Organization, 1957).

10. Ancel Keys and J. T. Anderson, *Symposium on Atherosclerosis: Proceedings* (Washington: National Academy of Sciences–National Research Council, 1954), quoted in Jacob Yerushalmy and Herman E. Hilleboe, "Fat in the Diet and Mortality from Heart Disease: A Methodological Note," *New York State Journal of Medicine* 57 (July 1957): 2344.

11. John Yudkin, *Pure, White and Deadly*, 2nd ed. (London: Penguin, 2012), p. 86; and J. Yudkin, "Diet and Coronary Thrombosis: Hypothesis and Fact," *Lancet* 2 (1957): 155.

12. Yudkin, *Pure, White and Deadly*, p. 63.

13. Edward H. Ahrens, "Dietary Control of Serum Lipids in Relation to Atherosclerosis," *Journal of the American Medical Association* 164 (August 24, 1957): 1905–1911, https://doi.org/10.1001/jama.1957.62980170017007d.

14. Carmel McCoubrey, "Edward Ahrens Cholesterol Researcher, Is Dead at 85," *New York Times*, December 16, 2000.

15. George V. Mann, "Diet-Heart: End of an Era," *New England Journal of Medicine* 297 (October 1977): 644–650, https://doi.org/10.1056/nejm197709222971206.

16. A. E. Bennett et al., "Sugar Consumption and Cigarette Smoking," *Lancet* 295 (May 16, 1970): 1011–1014, https://doi.org/10.1016/s0140-6736(70)91147-5.

17. Gerald Finking and Hartmut Hanke, "Nikolaj Nikolajewitsch Anitschkow (1885–1964) Established the Cholesterol-Fed Rabbit as a Model for Atherosclerosis Research," *Atherosclerosis* 135 (1997): 1–7, https://doi.org/10.1016/s0021-9150(97)00161-5.

18. Ancel Keys et al., "Effects of Diet on Blood Lipids in Man, Particularly Cholesterol and Lipoproteins," *Clinical Chemistry* 1 (1955): 34–52, http://clinchem .aaccjnls.org/content/clinchem/1/1/34.full.pdf.

19. U.S. Department of Agriculture and Department of Health and Human Services, *Scientific Report of the 2015 Dietary Guidelines Advisory Committee* (Washington: USDA and HHS, 2015), p. 17, https://health.gov/dietaryguidelines /2015-scientific-report/PDFs/Scientific-Report-of-the-2015-Dietary-Guidelines -Advisory-Committee.pdf.

20. Gary Taubes, "The Soft Science of Dietary Fat," *Science* 291 (March 30, 2001): 2536–45, https://doi.org/10.1126/science.291.5513.2536.

21. Ancel Keys, "Epidemiologic Aspects of Coronary Artery Disease," *Journal of Chronic Diseases* 6 (1957): 552–559.

22. Senate Select Committee, *Dietary Goals*, p. 3.

23. Manuel Hörl, "Fat: The New Health Paradigm," Credit Suisse (website), September 22, 2015, https://www.credit-suisse.com/corporate/en/articles/news -and-expertise/fat-the-new-health-paradigm-201509.html.

24. Study Group on Atherosclerosis and Ischaemic Heart Disease, *Proceedings*.

25. Ancel Keys, ed., *Seven Countries: A Multivariate Analysis of Death and Coronary Heart Disease* (Cambridge, MA: Harvard University Press, 1980), quoted by Teicholz in *Big Fat Surprise*, p. 38.

26. Study Group on Atherosclerosis and Ischaemic Heart Disease, *Proceedings*.

27. Alessandro Menotti et al., "Food Intake Patterns and 25-Year Mortality from Coronary Heart Disease: Cross-Cultural Correlations in the Seven Countries Study," *European Journal of Epidemiology* 15 (July 1999): 507–515.

28. Zoë Harcombe et al., "Evidence from Randomised Controlled Trials Did Not Support the Introduction of Dietary Fat Guidelines in 1977 and 1983: A Systematic Review and Meta-Analysis," *Open Heart* 2 (2015): 1–6, https://doi .org/10.1136/openhrt-2014-000196.

29. "Dietary Goals for the United States: Statement of the American Medical Association to the Select Committee on Nutrition and Human Needs, United States Senate," *Rhode Island Medical Journal* 60 (1977): 576–581.

30. Senate Select Committee, *Dietary Goals*, p. 10.

31. Harcombe et al., "Evidence from Randomised Controlled Trials."

32. For an account of the mass media's support for the government's message, see Teicholz, *Big Fat Surprise*, pp. 47–53.

33. Beth Schucker, "Change in Public Perspective on Cholesterol and Heart Disease," *Journal of the American Medical Association* 258 (December 1987): 3527, https://doi.org/10.1001/jama.1987.03400240059024.

34. Shi-Sheng Zhou et al., "B-Vitamin Consumption and the Prevalence of Diabetes and Obesity among the US Adults: Population Based Ecological Study," *BMC Public Health* 10, no. 1 (December 2, 2010): https://doi.org/10.1186/1471-2458-10-746.

35. Data from the USDA are reproduced in Roberto A. Ferdman, "The Generational Battle of Butter vs. Margarine," *Washington Post*, June 17, 2014.

36. D. Mozaffarian, A. Aro, and W. C. Willett, "Health Effects of Trans-Fatty Acids: Experimental and Observational Evidence," *European Journal of Clinical Nutrition* 63, no. S2 (May 2009): S5–S21, https://doi.org/10.1038/sj.ejcn.1602973.

37. World Health Organization, "The Top 10 Causes of Death," http://www.who.int/news-room/fact-sheets/detail/the-top-10-causes-of-death.

38. C. N. Hales and D. Barker, "Type 2 (Non-Insulin-Dependent) Diabetes Mellitus: The Thrifty Phenotype Hypothesis," *International Journal of Epidemiology* 42 (October 2013): 1215–1222, https://doi.org/10.1093/ije/dyt133.

39. Quoted in Caroline Fall and Clive Osmond, "David Barker," *Sight and Life* 27 (2013): 64–66.

40. Richard Weindruch and Rajindar S. Sohal, "Caloric Intake and Aging," *New England Journal of Medicine* 337 (October 2, 1997): 986–994, https://doi.org/10.1056/nejm199710023371407.

41. P. L. Hofman, "Insulin Resistance in Short Children with Intrauterine Growth Retardation," *Journal of Clinical Endocrinology and Metabolism* 82 (February 1997): 402–406, https://doi.org/10.1210/jc.82.2.402.

42. John N. Duggan and Anne E. Duggan, "The Possible Causes of the Pandemic of Peptic Ulcer in the Late 19th and Early 20th Century," *Medical Journal of Australia* 185 (December 2006): 667–669; John R. Brooks and Angelo J. Eraklis, "Factors Affecting the Mortality from Peptic Ulcer," *New England Journal of Medicine* 271 (October 15, 1964): 803–809, https://doi.org/10.1056/nejm196410152711601.

43. Gary Taubes, *Good Calories, Bad Calories: Challenging the Conventional Wisdom on Diet, Weight Control, and Disease* (New York: Alfred A. Knopf, 2007).

44. E. S. Gordon et al., "A New Concept in the Treatment of Obesity," *Journal of the American Medical Association* 186 (October 5, 1963): 50–60, https://doi.org/10.1001/jama.1963.63710010013014.

45. Robert J. Kuczmarski, "Increasing Prevalence of Overweight among U.S. Adults," *Journal of the American Medical Association* 272 (July 20, 1994): 205, https://doi.org/10.1001/jama.1994.03520030047027; Katherine M. Flegal, "Prevalence and Trends in Obesity among U.S. Adults, 1999–2000," *Journal of the American Medical Association* 288 (October 9, 2002): 1723, https://doi.org/10.1001/jama.288.14.1723; Centers for Disease Control and Prevention, *National Diabetes Statistics Report, 2017* (Atlanta: Centers for Disease Control and Prevention, 2017).

46. Alan W. Barclay and Jennie Brand-Miller, "The Australian Paradox: A Substantial Decline in Sugar Intake over the Same Timeframe That Overweight and Obesity Have Increased," *Nutrition* 3 (April 3, 2011): 491–504.

47. GBD 2015 Obesity Collaborators, "Health Effects of Overweight and Obesity in 195 Countries over 25 Years," *New England Journal of Medicine* 377 (July 6, 2017): 13–27, https://doi.org/10.1056/nejmoa1614362.

48. National Institute on Aging, "Global Aging," 2015, https://www.nia.nih.gov/research/publication/global-health-and-aging/living-longer.

49. Katherine M. Flegal et al., "Association of All-Cause Mortality with Overweight and Obesity Using Standard Body Mass Index Categories," *Journal of the American Medical Association* 309, no. 1 (January 2, 2013): 71–82, https://doi.org/10.1001/jama.2012.113905.

50. Frank M. Sacks et al., "Dietary Fats and Cardiovascular Disease: A Presidential Advisory from the American Heart Association," *Circulation* 136 (June 15, 2017): https://doi.org/10.1161/cir.0000000000000510.

51. U.S. Department of Health and Human Services (HHS) and U.S. Department of Agriculture, *2015–2020 Dietary Guidelines for Americans*, 8th ed. (Washington: HHS, 2015), https://health.gov/dietaryguidelines/2015/guidelines/.

52. Michael Pollan, *In Defense of Food: An Eater's Manifesto* (New York: Penguin Books, 2008).

53. Thomas C. Campbell, "A Plant-Based Diet and Animal Protein: Questioning Dietary Fat and Considering Animal Protein as the Main Cause of Heart Disease," *Journal of Geriatric Cardiology* 14 (May 2017): 331–337, https://doi.org/10.11909/j.issn.1671-5411.2017.05.011.

54. Ambika Satija et al., "Healthful and Unhealthful Plant-Based Diets and the Risk of Coronary Heart Disease in U.S. Adults," *Journal of the American College of Cardiology* 70, no. 4 (July 25, 2017): 411–22, https://doi.org/10.1016/j.jacc.2017.05.047.

55. Mayo Clinic, "DASH Diet: Healthy Eating to Lower Your Blood Pressure," April 8, 2016, http://www.mayoclinic.org/healthy-lifestyle/nutrition-and-healthy-eating/in-depth/dash-diet/art-20048456?pg=1.

56. Tomomi Shiiya et al., "Plasma Ghrelin Levels in Lean and Obese Humans and the Effect of Glucose on Ghrelin Secretion," *Journal of Clinical Endocrinology and Metabolism* 87 (January 2002): 240–244, https://doi.org/10.1210/jcem.87.1.8129.

57. Mancur Olson, *The Logic of Collective Action* (Cambridge, MA: Harvard University Press, 1965).

58. Quoted in Gerald M. Oppenheimer and I. Daniel Benrubi, "McGovern's Senate Select Committee on Nutrition and Human Needs versus the Meat Industry on the Diet-Heart Question (1976–1977)," *American Journal of Public Health* 104 (January 2014): 59–69, https://doi.org/10.2105/ajph.2013.301464.

59. Marion Nestle, *Food Politics: How the Food Industry Influences Nutrition and Health* (Berkeley: University of California Press, 2013), p. 3.

60. Jane E. Brody, "What's New in the Dietary Guidelines," *New York Times*, January 18, 2016.

61. Maira Bes-Rastrollo et al., "Financial Conflicts of Interest and Reporting Bias regarding the Association between Sugar-Sweetened Beverages and Weight Gain: A Systematic Review of Systematic Reviews," *PLoS Medicine* 10 (December 31, 2013): 1–9, https://doi.org/10.1371/journal.pmed.1001578.

62. Cristin E. Kearns, Laura A. Schmidt, and Stanton A. Glantz, "Sugar Industry and Coronary Heart Disease Research," *JAMA Internal Medicine* 176 (November 01, 2016): 1680–1685, https://doi.org/10.1001/jamainternmed.2016.5394.

63. P. Whoriskey, "Congress: We Need to Review the Dietary Guidelines for Americans," *Washington Post*, December 18, 2015.

Chapter 3

1. D. A. McCarron et al., "Normal Range of Human Dietary Sodium Intake: A Perspective Based on 24-Hour Urinary Sodium Excretion Worldwide," *American Journal of Hypertension* 26 (2013): 1218–1223.

2. B. Folkow and D. Ely, "Importance of the Blood Pressure–Heart Rate Relationship," *Blood Pressure* 7 (1998): 133–138.

3. A. M. Bernstein and W. C. Willett, "Trends in 24-h Urinary Sodium Excretion in the United States, 1957–2003: A Systematic Review," *American Journal of Clinical Nutrition* 92 (2010): 1172–1180; C. M. Pfeiffer et al., "Urine Sodium Excretion Increased Slightly among U.S. Adults between 1988 and 2010," *Journal of Nutrition* 144 (2014): 698–705.

4. D. A. McCarron et al., "Blood Pressure and Metabolic Responses to Moderate Sodium Restriction in Isradipine-Treated Hypertensive Patients," *American Journal of Hypertension* 10 (1997): 68–76.

5. The Hebrew Bible describes a "covenant of salt," an agreement requiring that sacrifices of both grain and animal were to be salted. Hawaiian lore holds that salt is a gift from the goddess Pele and serves as a symbol of good luck.

6. For more on this, see the renin-angiotensin-aldosterone system (RAAS) and atrial natriuretic peptide. See Michelle Minton, "Shaking up the Conventional Wisdom on Salt," Competitive Enterprise Institute Issue Analysis 2017 no. 1, January 2017, p. 7, for a description of the roles RAAS and ANP play.

7. L. Ambard and E. Beaujard, "Causes de l'hypertension arterielle," *General Archives of Medicine* 1 (1904): 520–533.

8. James DiNicolantonio, *The Salt Fix* (New York: Harmony Books, 2017).

9. C. B. Chapman and T. B. Gibbons, "The Diet and Hypertension: A Review," *Medicine (Baltimore)* 29, no. 1 (1950): 29–69; C. Lowenstein, "Über Beziehungen zwischen Kochsalzhaushalt und Blutdruck bei Nierenkranken," *Archiv für experimentelle Pathologie und Pharmakologie* 57 (1907): 137–161.

10. Kempner did not track potassium, but subsequent analysis of the Kempner diet found levels of up to 1.9 grams. See Chapman and Gibbons, "Diet and Hypertension."

11. W. Kempner, "Treatment of Hypertensive Vascular Disease with Rice Diet," *American Journal of Medicine* 4 (1948): 545–577.

12. Chapman and Gibbons, "Diet and Hypertension."

13. D. H. Ellison, "Physiology and Pathophysiology of Diuretic Action," in *Seldin and Giebisch's The Kidney: Physiology and Pathophysiology*, 4th ed., ed. R. J. Halpern, M. J. Caplan, and O. W. Moe (New York: Elsevier, 2008).

14. H. A. Schroeder et al., "Low Sodium Chloride Diets in Hypertension," *Journal of the American Medical Association (JAMA)* 140 (1949): 458–463.

15. Chapman and Gibbons, "Diet and Hypertension," p. 49.

16. H. Chasis et al., "Salt and Protein Restriction, Effects on Blood Pressure and Renal Hemodynamics in Hypertensive Patients," *JAMA* 142 (1950): 711–715.

17. L. K. Dahl and R. A. Love, "Evidence for Relationship between Sodium (Chloride) Intake and Human Essential Hypertension," *AMA Archives of Internal Medicine* 94 (1954): 525–531.

18. B. Joe, "Dr. Lewis Kitchener Dahl, the Dahl Rats and the 'Inconvenient Truth' about the Genetics of Hypertension," *Hypertension* 65 (2016): 963–969.

19. L. K. Dahl, "Possible Role of Salt Intake in the Development of Essential Hypertension," in *Essential Hypertension—An International Symposium* (Berlin: Springer, 1960), pp. 53–65.

20. L. K. Dahl, "Possible Role of Salt in Essential Hypertension," *International Journal of Epidemiology* 34, no. 2 (October 2005): 967–972, fig. 1.

21. Tyler Vigen, "Spurious Correlations," Tyler Vigen.com (website), http://www.tylervigen.com/spurious-correlations.

22. Dahl and Love, "Evidence for Relationship between Sodium and Hypertension."

23. Dahl, "Possible Role of Salt in Essential Hypertension," p. 970.

24. F. O. Simpson, "Salt and Hypertension: A Skeptical Review of the Evidence," *Clinical Science* 57 (1979): 463–480.

25. Simpson, "Salt and Hypertension."

26. T. R. Dawber et al., "Environmental Factors in Hypertension," in *The Epidemiology of Hypertension*, ed. J. Stamler (New York: Grune & Straton, 1967), pp. 255–288.

27. "Low-Sodium Diet Might Not Lower Blood Pressure," EurekAlert (website), press release, April 25, 2017.

28. Dahl, "Possible Role of Salt."

29. Dahl.

30. The term, cleverly borrowed from toxicology, refers to the amount of poison that would kill the average person. According to N. Trautmann, the actual lethal dose of sodium chloride—the acute amount that would be deadly for the average person—is around 3 grams per kilogram of body weight (*Assessing Toxic Risk* [Student Edition] [Arlington, VA: NSTA Press, 2001]). Translated, this means that for a person weighing 150 pounds, the lethal dose is around 205 grams of salt or just shy of 80,000 milligrams of sodium.

31. They used this same process for rats with the least blood pressure response to salt. The two strains are now known as Dahl salt-sensitive and Dahl salt-resistant rats and are still used in research to this day.

32. Dahl, "Possible Role of Salt."

33. L. K. Dahl et al. "Effects of Chronic Excess Salt Ingestion," *Circulation Research* 22 (1968): 605–617.

34. Simpson, "Salt and Hypertension," p. 464s.

35. J. D. Swales, "Dietary Salt and Hypertension," *The Lancet* 315 (1980): 1177.

36. J. T. Bennett and T. H. DiLorenzo, *From Pathology to Politics: Public Health in America* (New Brunswick, NJ: Transaction, 2000).

37. L. K. Dahl, "Salt in Processed Baby Foods," *American Journal of Clinical Nutrition* 21 (1968): 787–792.

38. Dahl, "Salt in Processed Baby Foods," p. 791.

39. Charles U. Lowe, Testimony in Hearings before the Select Committee on Nutrition and Human Needs, Part 13A—Nutrition and Private Industry, U.S. Senate, 90th Cong., 2nd sess., July 15–18, 1969, https://babel.hathitrust.org/cgi/pt?id=uc1.b3603570&view=1up&seq=7.

40. H. A. Schroeder, "Questions and Answers: Salt, Baby Food, and Hypertension," *JAMA* 186, no. 3 (1963): 279–283.

41. Dahl, Testimony in Hearings before the Select Committee on Nutrition and Human Needs.

42. Jean Mayer, "Hypertension, Salt Intake, and the Infant," *Postgraduate Medical Journal* 45 (1969): 229–230.

43. Jean Mayer, *Final Report*, White House Conference on Food, Nutrition, and Health (Washington: U.S. Government Printing Office, 1969), p. 44.

44. Dahl, "Salt in Processed Baby Foods," p. 789.

45. R. A. McCance, "Medical Problems in Mineral Metabolism," *The Lancet* 227 (1963): 765–768.

46. Gary Taubes, "The (Political) Science of Salt," *Science* 281 (1998): 898.

47. It is characteristic of many issues discussed in this book that the proponents of regulation claim a moral high ground and signal virtue, while attacking those who disagree in a particularly vicious ad hominem fashion.

48. Simpson, "Salt and Hypertension," p. 463s.

49. R. Cooper et al., "Urinary Sodium Excretion and Blood Pressure in Children: Absence of a Reproducible Association," *Hypertension* 5 (1983): 135–139.

50. S. Ljungman et al., "Sodium Excretion and Blood Pressure," *Hypertension* 3 (1981): 325.

51. G. MacGregor, "Sodium Is More Important than Calcium in Essential Hypertension," *Hypertension* 7 (1985): 628.

52. Simpson, "Salt and Hypertension," p. 466s.

53. Taubes, "The (Political) Science of Salt," p. 899.

54. A. R. Feinstein, "Tempest in a *P*-Pot," *Hypertension* 7 (1985): 318.

55. Taubes, "The (Political) Science of Salt," p. 898.

56. J. J. Filer Jr., "Salt in Infant Foods," *Nutrition Reviews* 29 (1971): 27–52.

57. M. J. Segal, "The Politics of Salt," in H. M. Sapolsky, *Consuming Fears: The Politics of Product Risk* (New York: Basic Books, 1986).

58. Julia Benson, "Nathan Pritikin, the Nutritionist Who Committed Suicide Is Eulogized," United Press International, February 28, 1985; Denise Minger, *Death by Food Pyramid: How Shoddy Science, Sketchy Politics and Shady Special Interests Have Ruined Our Health* (Malibu, CA: Primal Blueprint, 2014); Edward J. Boyer, "'Fellow Crusader' George McGovern at Rites: Pritikin Eulogized as a Bold Pioneer," *Los Angeles Times*, March 1, 1985.

59. Hearings before the Select Committee on Nutrition and Human Needs, Nutrition and Private Industry, U.S. Senate, 91st Cong., 1st sess., 1969, pp. 3983–4014, https://hdl.handle.net/2027/mdp.35112202636322?urlappend=%3Bseq=107.

60. Senate Hearings on Nutrition and Human Needs, Nutrition and Private Industry, p. 4099, https://hdl.handle.net/2027/umn.31951d021206832?urlappend=%3Bseq=227.

61. Select Committee on Nutrition and Human Needs, *Dietary Goals for the United States*, U.S. Senate (Washington: Government Printing Office, 1977), Internet Archive, https://archive.org/details/CAT10527234.

62. Gary Taubes, *Good Calories, Bad Calories* (New York: Alfred A. Knopf, 2007).

63. A. E. Harper, "Dietary Goals—A Skeptical View," *American Journal of Clinical Nutrition* 31 (1978): 310–321.

64. G. Taubes, "The Soft Science of Dietary Fat," *Science* 291 (2001): 2536–2545.

65. Select Committee on Nutrition and Human Needs, Dietary Goals for the United States, Supplemental Views, Statement of the American Medical Association to Senator McGovern, U.S. Senate, 95th Cong., 1st sess., 1977, p. 670, https://hdl.handle.net/2027/umn.31951d00283417h?urlappend=%3Bseq=678.

66. Alfred E. Harper, "Dietary Goals—A Skeptical View," *American Journal of Clinical Nutrition* 31, no. 2 (1978): 310–321.

67. Testimony before the Select Committee on Nutrition and Human Needs, "Diet Related to Killer Diseases, VII," U.S. Senate, 95th Cong., 1st sess., 1977, https://hdl.handle.net/2027/mdp.39015078681361?urlappend=%3Bseq=37.

68. Testimony before the Senate Committee on Nutrition and Human Needs, "Nutrition and Private Industry," U.S. Senate, 91st Cong., 1st sess., 1969, pp. 3983–4014, https://hdl.handle.net/2027/mdp.35112202636322?urlappend=%3Bseq=107.

69. Gilbert A. Leveille, Testimony before Committee on Agriculture, Nutrition, and Forestry, Subcommittee on Nutrition, Nutrition Research, U.S. Senate, 95th Cong., 2nd sess., 1978, p. 112, http://bit.ly/2BkffeF.

70. Mark Hegsted, Testimony before the Select Committee on Nutrition and Human Needs, "Diet Related to Killer Diseases, I," U.S. Senate, 94th Cong., 2nd sess., 1976, p. 209, https://hdl.handle.net/2027/pst.000014141805?urlappend=%3Bseq=215; Henry Blackburn, interview with Mark Hegsted, October 24, 2005, http://www.foodpolitics.com/wp-content/uploads/Hegsted.pdf; Letter from Mark Hegsted to Henry Blackburn, University of Minnesota, http://www.epi.umn.edu/cvdepi/document/hegsteds-personal-history-of-the-dietary-guidelines/.

71. Testimony before Senate Committee, "Diet Related to Killer Diseases, III," p. 57, https://hdl.handle.net/2027/mdp.39015078681320?urlappend=%3Bseq=17.

72. Robert Olson, Testimony before Select Committee on Nutrition and Human Needs, "Diet Related to Killer Diseases, VI," U.S. Senate, 95th Cong., 1st sess., 1977, pp. 9–10, https://hdl.handle.net/2027/umn.31951d00283427e?urlappend=%3Bseq=1.

73. Select Committee on Nutrition and Human Needs, "Dietary Goals for the United States, Supplemental Views," U.S. Senate, 95th Cong., 1st sess., 1977, p. 41, https://hdl.handle.net/2027/umn.31951d00283417h?urlappend=%3Bseq=1.

74. Testimony before the U.S. Senate Select Committee on Nutrition and Human Needs, "Diet Related to Killer Disease, I," 94th Cong., 2nd sess., 1976, p. 209, https://hdl.handle.net/2027/pst.000014141805?urlappend=%3Bseq=215.

75. Hearings before the Select Committee on Nutrition and Human Needs, 95th Cong., 1st sess., February 1 and 2, 1977, p. 296, https://hdl.handle.net/2027/mdp.39015078681304?urlappend=%3Bseq=1.

76. R. B. Sleeth and O. E. Kolari, "Sodium: Dietary and Labeling Issues," *Activities Report of the R and D Associates* 5 (1983): 167–182.

77. Committee on Agriculture, Nutrition, and Forestry, Subcommittee on Nutrition, "Nutrition Labeling and Information," U.S. Senate, 96th Cong., 2nd sess., 1980, p. 93, https://hdl.handle.net/2027/umn.31951d00291156n?urlappend=%3Bseq=1.

78. Hearings before the Select Committee on Nutrition and Human Needs of the United States Senate, "Diet Related to Killer Diseases, II," 95th Cong., 1st sess., February 1 and 2, 1977, p. 18, https://hdl.handle.net/2027/umn.31951d002834220?urlappend=%3Bseq=24. See also Hearings before the Subcommittee on Investigations and Oversight of the Committee on Science and Technology, "Sodium

in Food and High Blood Pressure," U.S. House of Representatives, 97th Cong., 1st sess., April 13 and 14, 1981, p. 16, https://babel.hathitrust.org/cgi/pt?id=mdp .39015082337414;view=1up;seq=20.

79. Harper, "Dietary Goals—A Skeptical View"; Food and Nutrition Board, Washington, National Research Council National Academy of Sciences, 1980; *Recommended Dietary Allowances, 9th edition* (Washington: U.S. National Research Council Food and Nutrition Board).

80. Segal, "The Politics of Salt."

81. Segal, p. 91.

82. U.S. Congress, House of Representatives, Committee on Science and Technology, Subcommittee on Investigations and Oversight, "Sodium and Blood Pressure," 97th Cong., 1st sess., 1981.

83. Rep. Albert Gore, Hearings before the Committee on Science and Technology, Subcommittee on Investigations and Oversight, Opening Statement, "Sodium and Blood Pressure," U.S. House of Representatives, 97th Cong., 1st sess., 1981.

84. John D. Laragh, Testimony before the House Committee on Science and Technology, "Sodium and Blood Pressure."

85. Laragh, "Sodium and Blood Pressure."

86. Gore, "Sodium and Blood Pressure." This is precisely the same argument Gore repeatedly marshalled in his global warming hearings.

87. Laragh, "Sodium and Blood Pressure."

88. Again, this tactic is reminiscent of how Gore repeatedly dangled the prospect of increasing global warming funds for NASA.

89. Michael Jacobson, Testimony before the House Committee on Science and Technology, "Sodium and Blood Pressure."

90. Gore, House Committee on Science and Technology Hearings, Closing Statement, "Sodium and Blood Pressure"; Segal, "The Politics of Salt."

91. Bill to amend the Federal Food, Drug, and Cosmetic Act, H.R. 4031, 97th Cong. (1982–1982), https://www.congress.gov/bill/97th-congress/house -bill/4031/actions. Although sodium-specific statutes failed to win approval in Congress, it did enact the Nutrition Labeling and Education Act in 1990, which granted FDA the power to require that manufacturers list the content of many nutrients, including sodium, in their prepared foods.

92. Segal, "The Politics of Salt," p. 101.

93. Segal, p. 97.

94. American Heart Association, Report of the Nutrition Committee, "Rationale of the Diet-Heart Statement of the American Heart Association," *Arteriosclerosis, Thrombosis, and Vascular Biology* 2, no. 2 (1982): 187.

95. Segal, "The Politics of Salt," p. 109.

96. Swales, "Dietary Salt and Hypertension," p. 1179.

97. J. H. Laragh, "Dietary Sodium and Essential Hypertension: Some Myths, Hopes, and Truths," *Annals of Internal Medicine* 98 (1983): 740.

98. J. J. Brown et al., "Should Dietary Sodium Be Reduced? The Sceptics' Position," *Quarterly Journal of Medicine* n.s. 53 (1984): 427.

99. J. J. Brown et al., letter to the editor, "Salt and Hypertension," *The Lancet* 324, no. 8415 (December 8, 1984): 1333, https://doi.org/10.1016/S0140-6736(84)90834-1.

100. Simpson, "Salt and Hypertension," p. 463s.

101. H. de Wardener et al., letter to the editor, "Salt and Hypertension," *The Lancet* 324, no. 8404 (September 22, 1984): 688–689, https://doi.org/10.1016/S0140 -6736(84)91237-6.

102. Brown et al., letter to the editor.

103. De Wardener et al., letter to the editor.

104. D. L. Longworth et al., "Divergent Blood Pressure Responses during Short-Term Sodium Restriction in Hypertension," *Clinical Pharmacology and Therapeutics* 27 (1980): 544–546; E. D. Freis, "Salt, Volume and the Prevention of Hypertension," *Circulation* 53 (1976): 589–595; H. R. Brunner et al., "Essential Hypertension: Renin and Aldosterone, Heart Attack and Stroke," *New England Journal of Medicine* 286 (1972): 441–449.

105. Brown et al., letter to the editor.

106. Michael Jacobson, "The Deadly White Powder," *Mother Jones,* July 1978, p. 12.

107. P. Sabatier, S. Hunter, and S. McLaughlin, "The Devil Shift: Perceptions and Misperceptions of Opponents," *Western Political Quarterly* 40 (1987): 449–476.

108. Graham A. MacGregor and H. E. de Wardener, *Salt, Diet and Health: Neptune's Poisoned Chalice: The Origins of High Blood Pressure* (Cambridge: Cambridge University Press, 1998), p. 199.

109. Swales J, "Salt, Diet, and Health," *The Lancet* 353 (1999): 1710.

110. D. A. McCarron et al., "Blood Pressure and Nutrient Intake in the United States," *Science* 224 (1984): 1392–1398.

111. Jane E. Brody, "Personal Health," *New York Times,* August 22, 1984.

112. Letters, "Hypertension and Calcium, *Science* 226 (1984): 384–389.

113. Christine Russell, "Study Tying Lack of Salt to Hypertension Disputed," *Washington Post,* June 23, 1984.

114. The Intersalt study is discussed in detail below.

115. Goeffrey Cannon, *The Politics of Food* (London: Ebury Press, 1988), p. 299.

116. Taubes, "The (Political) Science of Salt," p. 899.

117. G. A. MacGregor and P. S. Sever, "Salt—Overwhelming Evidence but Still No Action: Can a Consensus Be Reached with the Food Industry?" *British Medical Journal* 312 (1996): 1288.

118. Matthew Sinclair, "Taxpayer Funded Lobbying and Political Campaigning," The Taxpayers' Alliance, August 3, 2009.

119. National Heart, Lung, and Blood Institute, *Primary Prevention of Hypertension: Clinical and Public Health Advisory from the National High Blood Pressure Education Program* (Washington: Department of Health and Human Services, November 2002).

120. W. C. Smith et al., "Urinary Electrolyte Excretion, Alcohol Consumption, and Blood Pressure in the Scottish Heart Health Study," *British Medical Journal* 297 (1988): 330.

121. J. D. Swales, "Salt Saga Continued," *British Medical Journal* 297 (1988): 307–308.

122. This result was later criticized because the study had not been originally designed to answer this question. This type of "data mining," as discussed in an early chapter of this book, is known as hypothesizing after the results are known, or HARKing. The authors, however, insisted they had always intended to investigate

this relationship and had just "stupidly" forgotten to include it in their original proposal. Taubes, "The (Political) Science of Salt."

123. Linda C. Higgins, "Big Study's Results Didn't Convert Critics to Blanket Salt Restriction," *Medical World News*, November 14, 1988.

124. William Ira Bennett, "Body and Mind; the Salt Alarm," *New York Times*, January 22, 1989.

125. D. Q. Haney, "Salt Talks: For 90 Percent of Americans, Salt Doesn't Matter Much," Associated Press, November 7, 1990.

126. Taubes, "The (Political) Science of Salt."

127. National High Blood Pressure Education Program, "Working Group Report on Primary Prevention of Hypertension," *Archives of Internal Medicine* 153 (1993): 186–208, https://hdl.handle.net/2027/uc1.31210010642625?urlappend=%3Bseq=3.

128. Taubes, "The (Political) Science of Salt."

129. Taubes.

130. P. Elliot et al., "Intersalt Revisited: Further Analyses of 24 Hour Sodium Excretion and Blood Pressure within and across Populations," *British Medical Journal* 312 (1996): 1249–1253.

131. M. Law, "Commentary: Evidence on Salt Is Consistent," *British Medical Journal* 312 (1996): 1284.

132. BMJ Letters, "Intersalt Data," *British Medical Journal* 315 (1997): 484–488.

133. C. S. Wiysonge et al., "Beta-Blockers for Hypertension," *Cochrane Database of Systematic Reviews* 15 (2012): CD002003.

134. B. Folkow, "Critical Review of Studies on Salt and Hypertension," *Clinical and Experimental Hypertension* 14 (1992): 2.

135. B. Folkow and D. Ely, "The Salt Problem Revisited: Importance of the Blood Pressure—Heart Rate Relationship," *Blood Pressure* 7 (1998): 133–138.

136. D. A. Freedman and D. B. Petitti, "Salt and Blood Pressure: Conventional Wisdom Reconsidered," *Evaluation Review* 25 (2001): 267–287.

137. The author reached out to Elliot for comment, but received no response.

138. See M. T. Jensen et al., "Resting Heart Rate Is Associated with Cardiovascular and All-Cause Mortality after Adjusting for Inflammatory Markers: The Copenhagen City Heart Study," *European Journal of Preventive Cardiology* 19 (2012): 102–108; J. E. Ho et al., "Long-Term Cardiovascular Risks Associated with an Elevated Heart Rate: The Framingham Heart Study," *Journal of the American Heart Association* 3 (2014): e000668; N. A. Graudal et al., "Reduced Dietary Sodium Intake Increases Heart Rate. A Meta-analysis of 63 Randomized Controlled Trials Including 72 Study Populations," *Frontiers in Physiology* 7 (2016): 111.

139. McCarron et al., "Normal Range of Human Dietary Sodium Intake."

140. P. Goldstein and M. Leshem, "Dietary Sodium, Added Salt, and Serum Sodium Associations with Growth and Depression in the U.S. General Population," *Appetite* 79 (2014): 83–90.

141. H. W. Cohen, S. M. Hailpern, and M. H. Alderman, "Sodium Intake and Mortality Follow-Up in the Third National Health and Nutrition Examination Survey (NHANES III)," *Journal of General Internal Medicine* 23 (2008): 1297–1302.

142. Q. Yang et al., "Sodium and Potassium Intake and Mortality among U.S. Adults: Prospective Data from the Third National Health and Nutrition Examination Survey," *Archives of Internal Medicine* 171 (2011): 1183–1191.

143. H. W. Cohen, M. Alderman, and N. Graudal, "Reanalysis of NHANES III Data on Sodium Association with Mortality: Appropriate Adjustment for Potassium Not Performed," *Archives of Internal Medicine* 171 (2011): 2063.

144. K. Stolarz-Skrzypek et al., "Fatal and Nonfatal Outcomes, Incidences of Hypertension, and Blood Pressure Changes in Relation to Urinary Sodium Excretion," *JAMA* 305 (2011): 1777–1785.

145. F. J. He et al., "Does Reducing Salt Intake Increase Cardiovascular Mortality?" *Kidney International* 80 (2011): 696–698.

146. "Salt and Cardiovascular Disease Mortality," Editorial, *The Lancet* 377 (2011): 1626.

147. M. J. O'Donnell et al., "Urinary Sodium and Potassium Excretion and Risk of Cardiovascular Events," *JAMA* 306 (2011): 2229–2238.

148. M. J. O'Donnell et al., "Urinary Sodium and Potassium Excretion, Mortality, and Cardiovascular Events," *New England Journal of Medicine* 371 (2014): 512–623.

149. N. Graudal et al., "Compared with Usual Sodium Intake Low- and Excessive-Sodium Diets Are Associated with Increased Mortality: A MetaAnalysis," *American Journal of Hypertension* 27 (2014): 1129–1137.

150. F. J. He and G. A. MacGregor, "Salt Intake and Mortality," *American Journal of Hypertension* 27 (2014): 1424.

151. A. Mente et al., "Associations of Urinary Sodium Excretion with Cardiovascular Events in Individuals with and without Hypertension: A Pooled Analysis of Data from Four Studies," *The Lancet* 388, no. 10043 (2016): 465–475.

152. R. P. Heaney, "Making Sense of the Science of Sodium," *Nutrition Today* 50, no. 2 (2015): 65.

153. F. P. Cappuccio, "Sodium and Cardiovascular Disease," *The Lancet* 388 (2016): 2112.

154. N. Graudal, "A Radical Sodium Reduction Policy Is Not Supported by Randomized Controlled Trials or Observational Studies: Grading the Evidence," *American Journal of Hypertension* 29 (2016): 543–548.

155. Lisa Nainggolan, "New Salt Paper Causes Controversy," *Medscape*, May 3, 2011, https://www.medscape.com/viewarticle/790910.

156. F. J. He, S. Pombo-Rodrigues, and G. A. MacGregor, "Salt Reduction in England from 2003 to 2011: Its Relationship to Blood Pressure, Stroke and Ischaemic Heart Disease Mortality," *BMJ Open* 4 (2012): e004549.

157. D. A. McCarron et al., "Can Dietary Sodium Intake Be Modified by Public Policy?" *Clinical Journal of the American Society of Nephrology* 4 (2009): 1878–1882.

158. T. Laatikainen et al., "Blood Pressure, Sodium Intake, and Hypertension Control: Lessons from the North Karelia Project," *Global Heart* 11 (2012): 191–199.

159. F. J. He and G. A. Macgregor, "A Comprehensive Review on Salt and Health and Current Experience of Worldwide Salt Reduction Programmes," *Journal of Human Hypertension* 23 (2008): 363–384. Differences between nations in the changing prevalence of cardiovascular disease could, of course, be related to numerous other factors, including smoking rates, alcohol use, changes in body mass index, and population age.

160. See also A. M. Bernstein and W. C. Willett, "Trends in 24-h Urinary Sodium Excretion in the United States, 1957–2003: A Systematic Review," *American Journal of Clinical Nutrition* 92 (2010): 1172–1180.

161. "Effects of Weight Loss and Sodium Reduction Intervention on Blood Pressure and Hypertension Incidence in Overweight People with High-Normal Blood Pressure: The Trials of Hypertension Prevention, Phase II," *Archives of Internal Medicine* 157 (1997): 657–667.

162. P. K. Whelton et al., "Sodium Reduction and Weight Loss in the Treatment of Hypertension in Older Persons: A Randomized Controlled Trial of Nonpharmacologic Interventions in the Elderly (TONE)," *JAMA* 279 (1998): 839–846.

163. J. Oliver, E. L. Cohen, and J. V. Neel, "Blood Pressure, Sodium Intake, and Sodium Related Hormones in the Yanomamo Indians, a 'No-Salt' Culture," *Circulation* 52 (1975): 146–151.

164. T. Yoshimoto and Y. Hirata, "Aldosterone as a Cardiovascular Risk Hormone," *Endocrine Journal* 54 (2007): 359–370; G. Rossi et al., "Aldosterone as a Cardiovascular Risk Factor," *Trends in Endocrinology and Metabolism* 16 (2005): 104–107.

165. N. Graudal, "Population Data on Blood Pressure and Dietary Sodium and Potassium Do Not Support Public Health Strategy to Reduce Salt Intake in Canadians," *Canadian Journal of Cardiology* 32 (2017): 283–285.

166. B. Folkow, "On Bias in Medical Research: Reflections on Present Salt-Cholesterol Controversies," *Scandinavian Cardiovascular Journal* 45 (2011): 196.

167. *Dietary Guidelines for Americans 2005* (Washington: Department of Health and Human Services and Department of Agriculture, 2005).

168. M. C. Blank, "Total Body Na+ Depletion without Hyponatremia Can Trigger Overtraining-like Symptoms with Sleeping Disorders and Increasing Blood Pressure: Explorative Case and Literature Study," *Medical Hypotheses* 79 (2012): 799–804.

169. M. Maillot, P. Monsivais, and A. Drewnowski, "Food Pattern Modeling Shows That the 2010 Dietary Guidelines for Sodium and Potassium Cannot Be Bet Simultaneously," *Nutrition Research* 33 (2013): 188–194.

170. M. M. Covelli, "A Review of Long-Term Effects of Low Sodium Diet versus High Sodium Diet on Blood Pressure, Renin, Aldosterone, Catecholamines, Cholesterol and Triglyceride," *Evidence-Based Nursing* 15 (2012): 70–71.

171. Shelly Wood, "Calling All Physicians: The Salt 'Debate' Must Stop," *Medscape*, June 18, 2014, https://www.medscape.com/viewarticle/826970.

172. L. Trinquart, D. M. Johns, and S. Galea, "Why Do We Think We Know What We Know? A Metaknowledge Analysis of the Salt Controversy," *International Journal of Epidemiology* 45 (2016): 252.

173. Michael Oakes, *Bad Foods: Changing Attitudes about What We Eat* (New York: Routledge, 2004), p. 86.

174. Folkow, "On Bias in Medical Research."

175. "Cholesterol: And Now the Bad News . . . ," *Time*, March 26, 1984.

176. "Cholesterol . . . and Now the Good News," *Time*, July 19, 1999.

177. M. Y. Yakoob et al., "Circulating Biomarkers of Dairy Fat and Risk of Incident Diabetes Mellitus among Men and Women in the United States in Two Large Prospective Cohorts," *Circulation* 133 (2016): 1645–1654.

178. *2015–2020 Dietary Guidelines for Americans* (Washington: Department of Health and Human Services and Department of Agriculture, 2015), https://health.gov/dietaryguidelines/2015/guidelines/.

179. Michelle Minton, "FDA's Trans Fat 'Ban' a First Foray into Controlling Americans' Diets," Competitive Enterprise Institute (blog), July 14, 2014.

180. M. Kratz, T. Baars, and S. Guyenet, "The Relationship between High-Fat Dairy Consumption and Obesity, Cardiovascular, and Metabolic Disease," *European Journal of Nutrition* 52 (2013): 1–24.

181. K. Harris et al., "Is the Gut Microbiota a New Factor Contributing to Obesity and Its Metabolic Disorders?" *Journal of Obesity* 2012 (January 24, 2012): https://doi.org/10.1155/2012/879151.

182. R. G. Walker, P. K. Whelton, and H. Saito, "Relation between Blood Pressure and Renin, Renin Substrate, Angiotensin II, Aldosterone and Urinary Sodium and Potassium in 574 Ambulatory Subjects," *Hypertension* 1 (1979): 287–291.

183. P. Strazzullo et al., "Salt Intake, Stroke, and Cardiovascular Disease: Meta-analysis of Prospective Studies," *British Medical Journal* (November 24, 2009): 339:b4567, https://doi.org/10.1136/bmj.b4567.

184. H. W. Cohen and M. H. Alderman, "Errors in Data and Questionable Methodology Impact Conclusions on Sodium and CVD," *British Medical Journal* (2009): 339:b4567, https://doi.org/10.1136/bmj.b4567.

185. Alderman (1995) found a trend of reduced stroke with increased potassium; Tunstall-Pedoe (1997) found "an unexpectedly powerful protective relation of dietary potassium to all-cause mortality"; Nagata (2004) found that "adjustment for potassium intake had a great impact on [hazard ratio] estimates"; Larsson (2008) found that "potassium intake was also inversely associated with risk of cerebral infarction"; and Umesawa (2008) found "potassium intake was inversely associated with mortality from coronary heart disease and total cardiovascular disease." M. H. Alderman et al., "Low Urinary Sodium Is Associated with Greater Risk of Myocardial Infarction Among Treated Hypertensive Men," *Hypertension* 25 (1995): 1144–1152; H. Tunstall-Pedoe, M. Woodward, R. Tavendale, R. A'Brook, and M. K. McCluskey, "Comparison of the Prediction by 27 Different Factors of Coronary Heart Disease and Death in Men and Women of the Scottish Heart Health Study: Cohort Study," *British Medical Journal* 315 (1997): 722; C. Nagata, N. Takatsuka, N. Shimizu, and H. Shimizu, "Sodium Intake and Risk of Death from Stroke in Japanese Men and Women," *Stroke* 35 (2004): 1545; S. C. Larsson, M. J. Virtanen, and M. Mars, "Magnesium, Calcium, Potassium, and Sodium Intakes and Risk of Stroke in Male Smokers," *Archives of Internal Medicine* 168 (2008): 463; M. Umesawa et al., "Relations between Dietary Sodium and Potassium Intakes and Mortality from Cardiovascular Disease: The Japan Collaborative Cohort Study for Evaluation of Cancer Risks," *American Journal of Clinical Nutrition* 88 (2008): 195.

186. O'Donnell et al., "Urinary Sodium and Potassium Excretion."

187. N. R. Cook, I. J. Appel, and P. K. Whelton, "Sodium Intake and All-Cause Mortality over 20 Years in the Trials of Hypertension Prevention," *Journal of the American College of Cardiology* 68 (2016): 1609–1617.

188. S. R. Smith, P. E. Klotman, and L. P. Svetkey, "Potassium Chloride Lowers Blood Pressure and Causes Natriuresis in Older Patients with Hypertension," *Journal of the American Society of Nephrology* 2, no. 8 (1992): 1302–1309; G. G. Krishna, "Effect of Potassium Intake on Blood Pressure," *Journal of the American Society of Nephrology* 1, no. 1 (1990): 43–52; F. M. Sacks et al., "Effect on Blood Pressure of Potassium, Calcium and Magnesium in Women with Low Habitual Intake," *Hypertension* 31, no. 1 (1998): 131–138.

189. J. Poorolajal et al., "Oral Potassium Supplementation for Management of Essential Hypertension: A Meta-analysis of Randomized Controlled Trials," *PLoS One* 12 (2017): e0174967.

190. J. M. Geleijnse, F. J. Kok, and D. E. Grobbee, "Blood Pressure Response to Changes in Sodium and Potassium Intake: A Metaregression Analysis of Randomised Trials," *Journal of Human Hypertension* 17 (2003): 471–480.

191. M. E. Cogswel et al., "Sodium and Potassium Intakes among U.S. Adults: NHANES 2003–2008," *American Journal of Clinical Nutrition* 96 (2012): 647–657.

192. H. Lee et al., "Potassium Intake and the Prevalence of Metabolic Syndrome: The Korean National Health and Nutrition Examination Survey, 2008–2010," *PLoS One* 8 (2013): e55106.

193. N. J. Aburto et al., "Effects of Increased Potassium Intake on Cardiovascular Risk Factors and Disease: Systematic Review and Meta-analysis," *British Medical Journal* 346 (2013): f1378.

194. Maillot, Monsivais, and Drewnowski, "Food Pattern Modeling."

195. S. L. Shuger et al., "Body Mass Index as a Predictor of Hypertension Incidence among Initially Healthy Normotensive Women," *American Journal of Hypertension* 16 (2008): 613–619, https://doi.org/10.1038/ajh.2008.169.

196. "Effects of Nonpharmacologic Interventions on Blood Pressure of Persons with High Normal Levels," *JAMA* 267 (1992): 1213–1220; "Effects of Weight Loss and Sodium Reduction Intervention on Blood Pressure and Hypertension Incidence in Overweight People with High-Normal Blood Pressure: The Trials of Hypertension Prevention, Phase II," *Archives of Internal Medicine* 157 (1997): 657–667, https://doi.org/10.1001/archinte.1997.00440270105009.

197. D. T. Lackland, "Racial Difference in Hypertension: Implications for High Blood Pressure Management," *American Journal of the Medical Sciences* 348 (2014): 135–138, https://doi.org/10.1097/MAJ.0000000000000308.

198. I. T. Meredith, G. I. Jennings, and M. D. Esler, "Time Course of the Antihypertensive and Autonomic Effects of Regular Exercise in Human Subjects," *Journal of Hypertension* 8 (1990): 859–866.

199. Christopher Snowdon, "Don't Believe the Outrage. Anti-Sugar Activists Have Scored a Colossal Triumph," *Spectator Health*, August 18, 2016.

200. H. Boeing et al., "Critical Review: Vegetables and Fruit in the Prevention of Chronic Diseases," *European Journal of Nutrition* 51 (2012): 637–663.

Chapter 4

1. M. J. Brownstein, "A Brief History of Opiates, Opioid Peptides, and Opioid Receptors," *Proceedings of the National Academy of Sciences* 90 (1993): 5391–5393, http://www.pnas.org/content/90/12/5391.short.

2. Stephen Kishner, "Opioid Equivalents and Conversions," *Medscape,* January 29, 2018, https://emedicine.medscape.com/article/2138678-overview.

3. Alistair D. Corbett et al., "75 Years of Opioid Research: The Exciting but Vain Quest for the Holy Grail," *British Journal of Pharmacology* 147, Supp. 1 (January 2009): S153–S162, https://www.ncbi.nlm.nih.gov/pmc/articles /PMC1760732/.

4. Emily O. Doumas and Gary J. Pollack, "Opioid Tolerance Development: A Pharmacokinetic/Pharmacodynamic Perspective," *AAPS Journal* 10, no. 4 (2008): 537, https://www.ncbi.nlm.nih.gov/pmc/articles/PMC2628209/; Gavril W. Pasternak and Ying-Xian Pang, "Mu Opioids and Their Receptors: Evolution of a Concept," *Pharmacological Reviews* 65, no. 4 (2013): 1257–1317, https://www.ncbi .nlm.nih.gov/pmc/articles/PMC3799236/.

5. "Hydrocodone," *Toxicology Data Network,* September 19 2018, https:// toxnet.nlm.nih.gov/cgi-bin/sis/search/a?dbs+hsdb:@term+@DOCNO+3097.

6. "Safety Data Sheet," Purdue Pharma, April 23, 2015, https://www.purdueph arma.com/wp-content/uploads/2015/07/OxyContin-Tablets-23-Apr-15.pdf.

7. Jeffrey A. Singer, "The Myth of an Opioid Prescription Crisis," Cato Policy Report, September/October 2017, https://www.cato.org/policy-report/septemberoctober -2017/myth-opioid-prescription-crisis.

8. M. Ersek, M. M. Cherrier, S. S. Overman, and G. A. Irving, "The Cognitive Effects of Opioids," *Pain Management Nursing* 5, no. 2 (2004): 75–93, https:// www.ncbi.nlm.nih.gov/pubmed/15297954; S. Dublin et al., "Prescription Opioids and Risk of Dementia or Cognitive Decline: A Prospective Cohort Study," *Journal of the American Geriatrics Society* 63, no. 8 (2015): 1519–1526, https://www.ncbi .nlm.nih.gov/pubmed/26289681.

9. Monica Bawor et al., "Testosterone Suppression in Opioid Users: A Systematic Review and Meta-analysis," *Drug and Alcohol Dependence* 149 (2015): 1–9.

10. "NSAIDs: Optimizing Pain Management through Risk Reduction," Supplement *American Journal of Managed Care* (2013), http://www.ajmc.com/journals /supplement/2013/a467_nov13_nsaid/a467_nov13_nalamachu_s261. See also FDA Drug Safety Communication, "FDA Strengthens Warning That Non-Aspirin NSAIDs Can Cause Heart Attacks or Strokes," July 2015.

11. Zachary A. Marcum and Joseph T. Hanlon, "Recognizing the Risks of Chronic Nonsteroidal Anti-Inflammatory Drug Use in Older Adults," *Pain Medicine* (2010): 24–27.

12. Centers for Disease Control and Prevention, "High Blood Pressure," February 13, 2019, https://www.cdc.gov/bloodpressure/index.htm.

13. Rose A. Rudd et al., *Increases in Drug and Opioid Overdose Deaths— United States, 2000–2014* (Atlanta: Centers for Disease Control and Prevention, 2016).

14. Marcella Saccò et al., "The Relationship between Blood Pressure and Pain," *Journal of Clinical Hypertension* 15, no. 8 (2013): 600–605, https://doi .org/10.1111/jch.12145. See also Raymond Sinatra, "Causes and Consequences of Inadequate Management of Acute Pain," *Pain Medicine* 11 (December 2010): 1859–1971.

15. Forest Tennant, "Treat the Pain . . . Save a Heart," *Practical Pain Management* 10, no. 8 (October 2010): https://www.practicalpainmanagement.com

/pain/other/co-morbidities/treat-pain-save-heart; Judith A Whitworth et al., "Cardiovascular Consequences of Cortisol Excess," *Vascular Health Risk Management* 1, no. 4 (December 2005): 291–299, https://www.ncbi.nlm.nih.gov/pmc/articles/PMC1993964/.

16. "ASCOT and Other Study Results Show Some Surprising 'Causes' of Hypertension: Chronic Pain Associated with Increased Prevalence of Hypertension," *Medscape*, 2005, https://www.medscape.org/viewarticle/501272_5.

17. "Ibuprofen," *Toxicology Data Network*, March 21, 2005, https://toxnet.nlm.nih.gov/cgi-bin/sis/search/a?dbs+hsdb:@term+@DOCNO+3099.

18. "Naproxen," National Center for Biotechnology Information, PubChem Database, June 22, 2019, https://pubchem.ncbi.nlm.nih.gov/compound/Naproxen#section=Interactions.

19. Laura J. Hunter, David M. Wood, and Paul I. Dargan, "The Patterns of Toxicity and Management of Acute Nonsteroidal Anti-inflammatory Drug (NSAID) Overdose," *Open Access Emergency Medicine* (2011): 39–48.

20. Chelsea Carmona, "What Opioid Hysteria Leaves Out: Most Overdoses Involve a Mix of Drugs," *The Guardian*, June 8, 2016, https://www.theguardian.com/us-news/commentisfree/2016/jun/08/opioid-epidemic-drug-mix-overdose-death.

21. NYC Health, *Unintentional Drug Poisoning (Overdose) Deaths in New York City* (New York: New York City Department of Health and Mental Hygiene, 2017).

22. Deborah Dowell, Tamara M. Haegerich, and Roger Chou, *CDC Guideline for Prescribing Opioids for Chronic Pain*, Centers for Disease Control and Prevention: Morbidity and Mortality Weekly Report, 2016.

23. Amy S. B. Bohnert et al., "Association between Opioid Prescribing Patterns and Opioid Overdose-Related Deaths," *Journal of the American Medical Association* (2011): 1315–1321.

24. N. Dasgupta et al., "Cohort Study of the Impact of High-Dose Opioid Analgesics on Overdose Mortality," *PubMed* 17, no. 1 (January 2016): 85–98, https://doi.org/10.1111/pme.12907.

25. NIDA, "The Neurobiology of Drug Addiction," January, 2007, https://www.drugabuse.gov/publications/teaching-packets/neurobiology-drug-addiction/section-iii-action-heroin-morphine/10-addiction-vs-dependence.

26. "How to Tell the Difference between Dependence vs. Addiction," The Recovery Village (blog), January 22, 2019, https://www.therecoveryvillage.com/recovery-blog/dependence-addiction/#gref.

27. Jacob Sullum, "'Opioid Epidemic' Myths," *Reason* (Online) May 18, 2016, http://reason.com/archives/2016/05/18/opioid-epidemic-myths.

28. Castlight Health, "New Study Reveals 32 Percent of Total Opioid Prescriptions Are Being Abused," press release, 2016, Castlight.

29. M. Noble et al., "Opioids for Long-term Treatment of Noncancer Pain," January 20, 2010, http://www.cochrane.org/CD006605/SYMPT_opioids-long-term-treatment-noncancer-pain.

30. Silvia Minozzi et al., "Development of Dependence Following Treatment with Opioid Analgesics for Pain Relief: a Systematic Review," *Addiction*

108, no. 4 (2012), https://onlinelibrary.wiley.com/doi/abs/10.1111/j.1360-0443.2012.04005.x.

31. K. E. Vowles et al., "Rates of Opioid Misuse, Abuse, and Addiction in Chronic Pain: A Systematic Review and Data Wynthesis," *Pain* (2015): 569–576.

32. Gabriel A. Brat et al., "Postsurgical Prescriptions for Opioid Naive Patients and Association with Overdose and Misuse: Retrospective Cohort Study," *British Medical Journal* (2017): 360:j5790; https://doi.org/10.1136/bmj.j5790.

33. Richard M. Marks and Edward J. Sachar, "Undertreatment of Medical In-patients with Narcotic Analgesics," *Annals of Internal Medicine* 78, no. 2 (1973): 173–181, http://annals.org/aim/article-abstract/687199/undertreatment-medical-inpatients-narcotic-analgesics.

34. Jane Porter, "Addiction Rare in Patients Treated with Narcotics," *New England Journal of Medicine* (January 10, 1980), https://doi.org/10.1056/NEJM1980 01103020221; Taylor Haney, "Doctor Who Wrote 1980 Letter on Painkillers Regrets That It Fed the Opioid Crisis," NPR.org, June 16, 2017, http://www.npr.org/sections/health-shots/2017/06/16/533060031/doctor-who-wrote-1980-letter-on-painkillers-regrets-that-it-fed-the-opioid-crisi.

35. Hershel Jick, Olli S. Miettinen, and Samuel Shapiro, "Comprehensive Drug Surveillance," *Journal of the American Medical Association* (1970): 1455–1460; R. R. Miller and H. Jick, "Clinical Effects of Meperidine in Hospitalized Medical Patients," *Journal of Clinical Pharmacology* (1978): 180–189.

36. Jacob Sullum, "No Relief in Sight," *Reason* (Online), January 1997, http://reason.com/archives/1997/01/01/no-relief-in-sight/.

37. Meredith Noble et al., "Opioids for Long-term Treatment of Noncancer Pain," Cochrane, January 20, 2010, http://www.cochrane.org/CD006605/SYMPTopioids-long-term-treatment-noncancer-pain; Silvia Minozzi et al., "Development of Dependence Following Treatment with Opioid Analgesics for Pain Relief: a Systematic Review," *Addiction* 108, no. 4 (2012): 688–698, https://doi.org/10.1111/j.1360-0443.2012.04005.x.

38. Daniel Coleman, "Health: Patient Care; Physicians Said to Persist in Under-treating Pain and Ignoring the Evidence," *New York Times*, December 31, 1987.

39. Sullum, "No Relief in Sight."

40. Brian F. Mandell, "The Fifth Vital Sign: A Complex Story of Politics and Patient Care," *Cleveland Clinic Journal of Medicine* (2016), http://www.mdedge.com/ccjm/article/109138/drug-therapy/fifth-vital-sign-complex-story-politics-and-patient-care.

41. Sonia Moghe, "Opioid History: From 'Wonder Drug' to Abuse Epidemic," *CNN*, October 14, 2016, https://www.cnn.com/2016/05/12/health/opioid-addiction-history/.

42. Laura Santhanam, "Prescription Opioids Tripled between 1999 and 2015, CDC Says," PBS *NewsHour*, July 6, 2017, https://www.pbs.org/newshour/health/prescription-opioids-tripled-1999-2015-cdc-says.

43. CDC, "Drug Overdose Deaths," December 19, 2018, https://www.cdc.gov/drugoverdose/data/statedeaths.html.

44. Jeffrey A. Singer et al., "Today's Nonmedical Opioid Users Are Not Yesterday's Patients; Implications of Data Indicating Stable Rates of Nonmedical

Use and Pain Reliever Use Disorder," *Journal of Pain Research* 12 (2019): 617–620, doi:10.2147/JPR.S199750.

45. J. Westermeyer, "The Role of Cultural and Social Factors in the Cause of Addictive Disorders," *Psychiatric Clinics of North America* (1999): 253–273; Catherine Spooner and Kate Hetherington, *Social Determinants of Drug Use* (Sydney: National Drug and Alcohol Research Centre, 2004).

46. T. J. Cicero, M. S. Ellis, and Z. A. Kasper, "Increased Use of Heroin as an Initiating Opioid of Abuse," *Addicitive Behavior* (2017): 63–66.

47. Hawre Jalal et al., "Changing Dynamics of the Drug Overdose Epidemic in the United States from 1979 through 2016," *Science*, September 21, 2018.

48. Maggie Fox, "Opioid Crisis Started 40 Years Ago, Report Argues," *NBC News*, September 20, 2019, https://www.nbcnews.com/storyline/americas-heroin-epidemic/opioid-crisis-started-40-years-ago-report-argues-n911456.

49. Axel Bugge, "Drug Deaths on the Rise in Europe for Third Year: Report," Reuters, June 6, 2017, https://www.reuters.com/article/us-europe-drugs/drug-deaths-on-the-rise-in-europe-for-third-year-report-idUSKBN18X1Y4.

50. Kristine Phillips, "The Pill Mill Doctor Who Prescribed Thousands of Opioids and Billed Dead Patients," *Washington Post*, September 2016, https://www.washingtonpost.com/news/to-your-health/wp/2016/09/22/deceitful-pill-mill-doctor-who-prescribed-thousands-of-opioids-and-billed-dead-patients-settles-civil-lawsuit/?utm_term=.5c5ffb675455.

51. Kelly Foreman, "Pill Mills Reaching into Kentucky Communities, Touching Everyone," *Kentucky Law Enforcement* 11, no. 1 (2012): 19, https://docjt.ky.gov/Magazines/Issue%2041/files/assets/downloads/page0019.pdf.

52. "Prescription Drug Monitoring Frequently Asked Questions (FAQ)," http://www.pdmpassist.org/content/prescription-drug-monitoring-frequently-asked-questions-faq.

53. Young Hee Nam et al., "State Prescription Drug Monitoring Programs and Fatal Drug Overdoses," *Journal of Managed Care* (2017): 297–303; Justine Malatt, "The Effect of Prescription Drug Monitoring Programs on Opioid Prescriptions and Heroin Crime Rates," West Lafayette, IN, Purdue University Economics Department Working Paper no. 1292, 2017.

54. Dowell, Haegerich, and Chou, *CDC Guideline for Prescribing Opioids for Chronic Pain*.

55. Abby Goodnough, "Opioid Prescriptions Fall after 2010 Peak, C.D.C. Report Finds," *New York Times*, July 6, 2016, https://www.nytimes.com/2017/07/06/health/opioid-painkillers-prescriptions-united-states.html.

56. "DEA Proposes Cutting Production of Some Opioid Painkillers," Reuters, August 4, 2017, https://www.reuters.com/article/us-dea-opioids/dea-proposes-cutting-production-of-some-opioid-painkillers-idUSKBN1AK1ZH.

57. "FDA Takes Important Step to Increase the Development of, and Access to, Abuse-Deterrent Opioids," U.S. Food and Drug Administration, March 24, 2016, https://www.fda.gov/newsevents/newsroom/pressannouncements/ucm492237.htm.

58. T. J. Cicero and M. S. Ellis, "Abuse-Deterrent Formulations and the Prescription Opioid Abuse Epidemic in the United States: Lessons Learned from

OxyContin," *Journal of the American Medical Association Psychiatry* (2015): 424–430; Pamela Leece, Aaron M. Orkin, and Meldon Kahan, "Tamper-Resistant Drugs Cannot Solve the Opioid Crisis," *Canadian Medical Association Journal* (2015): 717–718; Singer, "The Myth of an Opioid Prescription Crisis."

59. William N. Evans et al., "How the Reformulation of OxyContin Ignited the Heroin Epidemic," Notre Dame University Working Paper, February 14, 2018, https://www3.nd.edu/~elieber/research/ELP.pdf.

60. National Institute on Drug Abuse, "Emerging Trends and Alerts," 2019, https://www.drugabuse.gov/drugs-abuse/emerging-trends-alerts.

61. Richard Harris, "Heroin Use Surges, Especially among Women and Whites," *NPR.org*, July 7, 2015, http://www.npr.org/sections/health-shots/2015/07/07/420874860/heroin-use-surges-especially-among-women-and-whites.

62. Kiersten Nuñez, "Police Say Residence in Utah Was Home to One of the Largest Fentanyl Operations in the U.S.," Fox 13 Now, May 31, 2017, http://fox13now.com/2017/05/31/police-say-residence-in-utah-was-home-to-one-of-the-largest-fentanyl-operations-in-the-u-s/.

63. Nicole Lewis et al., "Fentanyl Linked to Thousands of Urban Overdose Deaths," August 15, 2017, https://www.washingtonpost.com/graphics/2017/national/fentanyl-overdoses/?utm_term=.1f3b9a092613.

64. Sullum, "'Opioid Epidemic' Myths."

65. Michael A. Yokell et al., "Presentation of Prescription and Nonprescription Opioid Overdoses to US Emergency Departments," *Journal of the American Medical Association* 174, no. 12 (2014): 2034–2037, https://jamanetwork.com/journals/jamainternalmedicine/fullarticle/1918924; Maia Szalavitz, "Opioid Addiction Is a Huge Problem, but Pain Prescriptions Are Not the Cause," *Scientific American* (blog), May 10, 2016.

66. Sarah Frostenson, "Opioid Overdoses Are Climbing, but Prescription Pain-killers Aren't Driving Them Anymore," *Vox*, April 1, 2017, https://www.vox.com/science-and-health/2017/4/1/15115380/prescription-painkiller-heroin-deaths.

67. Holly Hedegaard et al., "Drug Overdose Deaths in the United States, 1999–2016," *Centers for Disease Control and Prevention, National Center for Health Statistics Brief*, no. 294 (2017), https://www.cdc.gov/nchs/products/databriefs/db294.htm.

68. Elinore McCance-Katz, Assistant Secretary for Mental Health and Substance Use, Substance Abuse and Mental Health Services Administration, HHS, Testimony before the Senate Subcommittee on Health, Education, Labor and Pensions (HELP), titled "The Federal Response to the Opioid Crisis," 115th Cong., 1st sess., October 5, 2017, https://www.help.senate.gov/imo/media/doc/HHS%20Opioids%20Statement.pdf.

69. "Overdose Death Rates," *National Institute on Drug Abuse*, January 2019, https://www.drugabuse.gov/related-topics/trends-statistics/overdose-death-rates.

70. Jeffrey A. Singer, "Opinion: As Fentanyl Floods the Streets, Don't Blame the Doctors, Patients," *Cincinnati Enquirer*, June 8, 2018, https://www.cincinnati.com/story/opinion/2018/06/08/opinion-fentanyl-floods-streets-dont-blame-doctors-patients/679133002/; Jeffrey A. Singer, "In the Opioid Crisis, Keep Your

Eyes on Heroin and Fentanyl," *National Review* (Online), October 25, 2017, http://www.nationalreview.com/article/453058/opioid-crisis-efforts-curtail-prescribing-are-backfiring.

71. Christopher Moraff, "Feds' Pill Crackdown Drives Pain Patients to Heroin," *Daily Beast*, April 13, 2017, https://www.thedailybeast.com/feds-pill-crackdown-drives-pain-patients-to-heroin.

72. "Overdose Death Rates," *National Institute on Drug Abuse*, January 2019, https://www.drugabuse.gov/related-topics/trends-statistics/overdose-death-rates.

Chapter 5

1. Quoted in Richard J. Bonnie and Charles H. Whitebread, *The Marijuana Conviction: A History of Marijuana Prohibition in the United States* (New York: Lindesmith Center, 1999), p. 251.

2. Bonnie and Whitebread, *The Marijuana Conviction*, p. 256.

3. Hearing on the Drug Enforcement Administration before the Subcommittee on Crime, Terrorism, and Homeland Security of the House Judiciary Committee, 112th Cong., 2nd sess., 2012, pp. 30–31, https://www.hsdl.org/?view&did=742919.

4. Carrie Johnson, "DEA Chief Michele Leonhart to Retire," *All Things Considered*, NPR, April 21, 2015.

5. Colby Itkowitz, "Rep. Jared Polis: DEA Chief 'Terrible' at Her Job," *Washington Post*, May 29, 2014.

6. Controlled Substances Act, 21 U.S.C. § 812(b)(1) (1970).

7. "Lists of: Scheduling Actions Controlled Substances Regulated Chemicals," U.S. Department of Justice, Drug Enforcement Administration, December 2018, https://www.deadiversion.usdoj.gov/schedules/orangebook/orangebook.pdf.

8. Carrie Johnson, "DEA Rejects Attempt to Loosen Federal Restrictions on Marijuana," *All Things Considered*, NPR, August 10, 2016.

9. That commission, called the Shafer Commission, proposed decriminalizing marijuana and treating it more like alcohol. *Marihuana: A Signal of Misunderstanding*, First Report of the National Commission on Marihuana and Drug Abuse, Government Printing Office, Washington, 1972.

10. Controlled Substances Act, 21 U.S.C. § 811(b) (1970).

11. Drug Abuse Control Amendments, 1970, Part I, Hearings before the Subcommittee on Public Health and Welfare, 91st Cong., 2nd sess., 1970, p. 84, https://catalog.hathitrust.org/Record/008466858; see also Drug Policy Alliance and MAPS, *The DEA: Four Decades of Impeding and Rejecting Science*, June 10, 2014, https://www.drugpolicy.org/sites/default/files/DPA-MAPS_DEA_Science_Final.pdf.

12. *NORML v. Ingersoll*, 497 F.2d 654 (DC Cir. 1974).

13. *NORML v. Drug Enforcement Administration*, 559 F.2d 735, p. 742 (DC Cir. 1977).

14. *NORML v. Drug Enforcement Administration*, p. 742.

15. *NORML v. Drug Enforcement Administration*, 1980 U.S. App. LEXIS 13099 (DC Cir. 1980) (unpublished disposition).

16. In the Matter of Marijuana Rescheduling Petition, Docket No. 86-22: Opinion and Recommended Ruling, Findings of Fact, Conclusions of Law and Decision of Administrative Law Judge (1988), http://ccguide.org/young88.php.

17. *Alliance for Cannabis Therapeutics v. Drug Enforcement Administration*, 930 F.2d 936, p. 940 (DC Cir. 1991) (citations omitted).

18. *Alliance for Cannabis Therapeutics*, p. 940.

19. United Nations Convention on Psychotropic Substances, "Final Act of the United National Conference for the Adoption of a Protocol on Psychotropic Substances," February 21, 1971, https://www.unodc.org/pdf/convention_1971_en.pdf.

20. Russ Juskalian, "Why It's Hard to Do Marijuana Research," *Newsweek*, November 3, 2010.

21. In the Matter of Lyle E. Craker, PhD, Opinion and Recommended Ruling, Findings of Fact, Conclusions of Law, and Decision of the Administrative Law Judge, Admin. Proc. File. No. 05-16 (February 12, 2007), http://www.maps.org/research-archive/mmj/ALJfindings.PDF.

22. In the Matter of Lyle E. Craker, p. 19.

23. Controlled Substances Act, 21 U.S.C. § 823(a)(1) (1970).

24. Applications to Become Registered under the Controlled Substances Act to Manufacture Marijuana to Supply Researchers in the United States, 81 Fed. Reg. 53846 (August 12, 2016) (to be codified at 21 C.F.R. pt. 1301).

25. In the Matter of Lyle E. Craker, p. 84.

26. See timeline for "Marijuana for Symptoms of PTSD in U.S. Veterans," at http://www.maps.org/research/mmj/marijuana-us.

27. Statement on the Adequacy of Marijuana Provided by NIDA for Phase 2 Clinical Trials for PTSD in Veterans, Press Release, March 17, 2017, MAPS.org, https://goo.gl/5oofbb.

28. "Timeline of MAPS Blocked Vaporizer Research," MAPS.org, https://maps.org/research/mmj/vaporizer-research-news-timeline.

29. Christopher Ingram, "Jeff Sessions Personally Asked Congress to Let Him Prosecute Medical-Marijuana Providers," *Washington Post*, June 13, 2017.

30. Matt Zapotosky and Devlin Barrett, "Justice Department at Odds with DEA on Marijuana Research, MS-13," *Washington Post*, August 15, 2017.

31. Zapotosky and Barrett, "Justice Department at Odds with DEA."

32. "Hatch, Harris Follow Up with Sessions, DOJ Regarding Medical Marijuana Research," Press Release, August 30, 2018, https://lawprofessors.typepad.com/marijuana_law/2018/04/senators-orrin-hatch-and-kamala-harris-write-to-ag-jeff-sessions-to-push-for-more-medical-marijuana-.html.

33. Sadie Gurman, "Marijuana-Research Applications Go Nowhere at Justice Department," *Wall Street Journal*, September 8, 2018.

34. Saba Hamedy, "Conservative Senator's Statement Laced with Marijuana References," CNN.com, September 13, 2017, http://cnn.it/2hT8E65.

35. Mike Adams, "Congress Approves Bill to Expand Medical Marijuana Research in U.S.," *National Post* (Canada), September 18, 2018.

36. 81 Fed. Reg. 53846 (August 12, 2016).

37. 81 Fed. Reg. 53846.

38. Chris Roberts, "DEA Will Let You Apply to Grow Cannabis—But There's a Catch," *Cannabis Now*, August 11, 2016.

39. 81 Fed. Reg. 53846.

40. Lyle E. Craker; Denial of Application, 74 Fed. Reg. 2101 (January 1, 2009).

41. Jayson Chesler and Alexa Ard, "Feds Limit Research on Marijuana for Medical Use," *USA Today*, August 18, 2015.

42. Shaunacy Ferro, "It's Incredibly Difficult to Study Medical Marijuana," *Business Insider*, August 12, 2013.

43. David Nutt, "Here's Why We Hear So Many False Claims about Cannabis," *International Centre for Science in Drug Policy* (blog), September 11, 2015. As an indication of the fractal nature of science/government interaction, substituting the words "global warming" for "drugs" yields a totally analogous paragraph [eds.].

44. "A MAPS History of MDMA," Multidisciplinary Association for Psychedelic Studies, http://www.maps.org/research-archive/mdma/mapsmdma.html.

45. Bruce Eisner, *Ecstasy: The MDMA Story*, 2nd ed. (Berkeley, CA: Ronin Publishing, 1994), p. 1.

46. Schedules of Controlled Substances Proposed Placement of 3,4-Methylenedioxymethamphetamine into Schedule 1, 49 Fed. Reg. 30210 (July 27, 1984), http://www.maps.org/research-archive/dea-mdma/pdf/0194.PDF.

47. Eisner, *Ecstasy*, pp. 7–8.

48. Jerry Adler and Pamela Abramson, "Getting High on 'Ecstasy,'" *Newsweek*, April 15, 1985.

49. Quoted in Jerome Beck and Marsha Rosenbaum, *Pursuit of Ecstasy: The MDMA Experience* (Albany: State University of New York Press, 1994), p. 20.

50. Eisner, *Ecstasy*, p. 19.

51. Eisner, p. 19.

52. Eisner, p. 20.

53. "Letter from Nathaniel Branden, Ph.D.," Multidisciplinary Association for Psychedelic Studies, August 23, 1983, http://www.maps.org/research-archive/dea-mdma/pdf/0173.PDF.

54. In the Matter of MDMA Scheduling, No. 84-48, May 22, 1986, https://goo.gl/4f7y4a.

55. In the Matter of MDMA Scheduling, pp. 14, 22.

56. In the Matter of MDMA Scheduling, p. 26.

57. *Grinspoon v. Drug Enforcement Administration*, 828 F.2d 881, p. 890 (1st Cir. 1987).

58. Schedules of Controlled Substances; Scheduling of 3,4-Methylenedioxymethamphetamine (MDMA) into Schedule I of the Controlled Substances Act; Remand, 53 Fed. Reg. 5156 (February 22, 1988).

59. Schedules of Controlled Substances; Scheduling of 3,4-Methylenedioxymethamphetamine (MDMA) into Schedule I; Remand.

60. Marsha Rosenbaum and Rick Doblin, "Why MDMA Should Not Have Been Made Illegal," in *The Drug Legalization Debate*, ed. James A. Inciardi (Thousand Oaks, CA: Sage, 1991), http://druglibrary.org/schaffer/lsd/rosenbaum.htm.

61. "A MAPS History of MDMA."

62. Michael Pollan, "The Trip Treatment," *New Yorker*, February 9, 2015.

63. David J. Nutt, Leslie A. King, and David E. Nichols, "Effects of Schedule I Drug Laws on Neuroscience Research and Treatment Innovation," *Nature Reviews Neuroscience* 14 (June 2013): 577–585.

64. David J. Nutt, *Drugs without the Hot Air* (Cambridge: UIT Cambridge Ltd., 2012), p. 25.

65. Nutt, *Drugs without the Hot Air*, p. 25.

66. German Lopez, "Ecstasy Could Be Available for Medical Use by 2021," *Vox*, November 30, 2016.

67. "FDA Grants Breakthrough Therapy Designation for MDMA-Assisted Psychotherapy for PTSD, Agrees on Special Protocol Assessment for Phase 3 Trials," Multidisciplinary Association for Psychedelic Studies, press release, August 26, 2017.

68. Lauren Gill, "MDMA for PTSD Therapy Enters Final Round of Trials, Could Be Approved in U.S. and Canada by 2021," Newsweek.com, Jan. 21, 2018.

69. "MDMA-Assisted Psychotherapy," Multidisciplinary Association for Psychedelic Studies, https://maps.org/research/mdma.

70. Pollan, "The Trip Treatment."

Chapter 6

1. M. Bowden, A. Crow, and T. Sullivan, *Pharmaceutical Achievers: The Human Face of Pharmaceutical Research* (Philadelphia: Chemical Heritage Press, 2003); J. Swann, *Academic Scientists and the Pharmaceutical Industry* (Baltimore: Johns Hopkins University Press, 1988).

2. P. Starr, *The Social Transformation of American Medicine* (New York: Harper, 1982), p. 514.

3. S. Lewis, *Arrowsmith* (New York: Harcourt Brace Jovanovich, 1925); B. Bernheim, *Medicine at the Crossroads*, (New York: William Morrow, 1939), p. 256.

4. A. Starr and M. Edwards, "Mitral Replacement: The Shielded Ball Valve Prosthesis," *Journal of Thoracic Cardiovascular Surgery* 42 (1961): 673–682.

5. D. Starr, *Blood: An Epic History of Medicine and Commerce* (New York: Alfred A Knopf, 1998), p. 441.

6. G. Zachary and G. Pascal, *The Endless Frontier: Vannevar Bush, Engineer of the American Century* (New York: Free Press, 1997).

7. V. Bush, *Science: The Endless Frontier*, A Report to the President by Vannevar Bush, Director of the Office of Scientific Research and Development (Washington: U.S. Government Printing Office, 1945).

8. V. Harden, *A Short History of the National Institutes of Health* (Bethesda, MD: Office of NIH History, 2012).

9. National Institutes of Health, Appropriations History by Institute/Center (1938 to present), https://officeofbudget.od.nih.gov/approp_hist.html.

10. The University of Texas Southwestern Medical Center, "UT Southwestern Medical Center: Mission and History, 1943–1959," 2013, http://www.utsouthwestern.edu/about-us/mission-history/1943-1959.html.

11. As noted in Chapter 2, in the postwar era, defense was the primary reason for the federalization of much of the entire scientific enterprise.

12. L. Thomas, *The Youngest Science: Notes of a Medicine Watcher* (New York: Penguin Books, 1983).

13. V. Narayanamurti and T. Odumosu, *Cycles of Invention and Discovery: Rethinking the Endless Frontier* (Cambridge, MA: Harvard University Press, 2016); D. Sarewitz, "Saving Science," *New Atlantis* 29 (Spring/Summer 2016): 4–40, http://www.thenewatlantic.com/publications/savings-science; T. Kealey, *The Economic Laws of Scientific Research* (New York: St. Martin's, 1996); T. Kealey, "The Case against Public Science," *Cato Unbound* (blog), August 5, 2013, http://www.cato-unbound.org/2013/08/05/terence-kealey/case-against-public -science.

14. D. Contopoulos-Ioannidis, E. Ntzani, and J. Ioannidis, "Translation of Highly Promising Basic Science Research into Clinical Applications," *American Journal of Medicine* 114 (2003): 477–484.

15. Patent Rights in Inventions Made with Federal Assistance, 35 U.S.C. Chapter 18, 1980, http://www.law.cornell.edu/uscode/text/35/part-II/chapter-18.

16. Joseph Allen, "Does University Patent Licensing Pay Off?" IPWatchDog .com, January 27, 2014.

17. Walter D. Valdivia, *University Start-ups: Critical for Improving Technology Transfer*, Brookings Institution Report, November 20, 2013, http://www.brookings .edu/research/university-start-ups-critical-for-improving-technology-transfer/.

18. B. Munos, "Lessons from 60 Years of Pharmaceutical Innovation," *Nature Reviews Drug Discovery* 8 (2009): 959–968.

19. J. Reichert and C. Milne, "Public and Private Sector Contributions to the Discovery and Development of "Impact" Drugs," *American Journal of Therapeutics* 9 (2002): 543–555; B. Zycher, J. DiMasi, and C. Milne, "Private Sector Contributions to Pharmaceutical Science: Thirty-five Summary Case Histories," *American Journal of Therapeutics* 17 (2010): 101–120; B. Sampat and F. Lichtenberg, "What Are the Respective Roles of the Public and Private Sectors in Pharmaceutical Innovation?" *Health Affairs* 30 (2011): 332–339; A. Stevens et al., "The Role of Public-Sector Research in the Discovery of Drugs and Vaccines," *New England Journal of Medicine* 364 (2011): 535–541; H. Linker et al., "Where Do New Medicines Originate from in the EU?" *Nature Reviews Drug Discovery* 13 (2014): 92–93.

20. National Institutes of Health, "Heart Disease," Fact Sheet, 2016, https:// report.nih.gov/nihfactsheets/ViewFactSheet.aspx?csid=96.

21. American Heart Association–American Stroke Association, "Heart Disease, Stroke and Research Statistics at-a-Glance," 2016, http://www.heart.org /idc/groups/ahamah-public/@wcm/@sop/@smd/documents/downloadable/ucm _480086.pdf.

22. C. W. Tsao and R. S. Vasan, "Cohort Profile: The Framingham Heart Study (FHS): Overview of Milestones in Cardiovascular Epidemiology," *International Journal of Epidemiology* 44 (2015): 1800–1813.

23. J. Le Fanu, *The Rise and Fall of Modern Medicine* (New York: Carroll and Graf, 1999).

24. D. Cutler et al., "The Value of Antihypertensive Drugs: A Perspective on Medical Innovation," *Health Affairs* 26 (2007): 97–110.

25. T. Stossel, "The Discovery of Statins," *Cell* 134 (2008): 903–905.

26. S. Luria and M. Human, "A Nonhereditary Host-induced Variation of Bacterial Viruses," *Journal of Bacteriology* 64 (1952): 557–569; J. Watson and F. Crick, "Molecular Structure of Nucleic Acids: A Structure for Deoxyribose

Nucleic Acid," *Nature* 171 (1953): 737–738; M. McCarty, *The Transforming Principle: Discovering That Genes Are Made of DNA* (Markham, Ontario: Commonwealth Fund, 1985).

27. J. C. Venter, *A Life Decoded: My Genome: My Life* (New York: Penguin, 2007); J. Shreeve, *The Genome War* (New York: Ballantine Books, 2005).

28. C. Sherwin and R. Isensen, *First Interim Report on Project Hindsight (Summary)* (Washington: Office of the Director of Defense Research and Engineering, 1966).

29. J. Comroe and R. Dripps, "Scientific Basis for the Support of Biomedical Science," *Science* 192 (1976): 105–111.

30. R. Smith, "Comroe and Dripps Revisited," *British Medical Journal* 295 (1987): 1404–1407.

31. A. Kesselheim, T. Tan, and J. Avorn, "The Roles of Academia, Rare Diseases, and Repurposing in the Development of the Most Transformative Drugs," *Health Affairs* 34 (2015): 286–293.

32. R. Chakravarthy et al., "Public- and Private-Sector Contributions to the Research and Development of the Most Transformational Drugs in the Past 25 Years: From Theory to Therapy," *Therapeutic Innovation and Regulatory Science* 50 (2016): 759–768.

33. R. Merton, "Priorities in Scientific Discovery," *American Sociological Review* 22 (1957): 635–659.

34. D. Boorstin, *The Discoverers* (New York: Random House, 1983).

35. H. Zuckerman and R. Merton, "Patterns of Evaluation in Science: Institutionalisation, Structure and Functions of the Referee System," *Minerva* 9 (1971): 66–100.

36. L. Fleck, *Genesis and Development of a Scientific Fact* (Chicago: University of Chicago Press, 1979); C. Begley, A. Buchan, and U. Dirnagl, "Institutions Must Do Their Part for Reproducibility," *Nature* 525 (2015): 25–27; P. Feyerabend, *The Tyranny of Science* (Cambridge: Polity, 2012); J. Ioannidis, "Why Most Published Research Findings Are False," *PLoS Medicine* 2 (2005): e124.

37. T. Kuhn, *The Structure of Scientific Revolutions*, 2nd ed. (Chicago: University of Chicago Press, 1970).

38. M. Polanyi, *Personal Knowledge: Towards a Post-Critical Philosophy* (Chicago: University of Chicago Press, 1958).

39. T. Cooper and A. Ainsberg, *Breakthrough: Elizabeth Hughes, the Discovery of Insulin, and the Making of a Medical Miracle* (New York: St. Martin's, 2010).

40. M. Van Regenmortel, "Reductionism and Complexity in Molecular Biology," *EMBO Reports* 5 (2004): 1016–1020.

41. H. Hesse, *The Glass Bead Game* (New York: Henry Holt, 1943).

42. J. Newman, *The Idea of a University* (South Bend, IN: Notre Dame University Press, 2003); C. Kerr, *The Uses of the University* (Cambridge, MA: Harvard University Press, 1963).

43. U.S. Census Bureau, *QuickFacts. United States* (Washington: Department of Commerce).

44. National Institutes of Health, *Budget*, http://www.nih.gov/about-nih/what-we-do/budget.

45. J. Wyngaarden, "The Clinical Investigator as an Endangered Species," *New England Journal of Medicine* 301 (1979): 1254–1259.

46. H. Moses III et al., "Financial Anatomy of Biomedical Research," *Journal of the American Medical Association* 294 (2005): 1333–1342; "Industry Profile 2016," PhRMA, http://www.phrma.org/report/industry-profile-2016.

47. A. Schafer, *The Vanishing Physician-Scientist?* (Ithaca, NY: Cornell University Press, 2009).

48. L. Brass et al., "Are MD-PhD Programs Meeting Their Goals? An Analysis of Career Choices Made by Graduates of 24 MD-PhD Programs," *Academic Medicine* 85 (2010): 692–701.

49. R. Barr, "Ro1 Teams and Grantee Age Trends in Grant Funding," *Inside NIA* (blog), April 22, 2015, http://www.nia.nih.gov/research/blog/2015/04/ro1-teams-and-grantee-age-trends-grant-funding.

50. T. Stossel, *Pharmaphobia: How the Conflict-of-Interest Myth Undermines American Medical Innovation* (Lanham, MD: Rowman and Littlefield, 2015).

51. K. Woolley et al., "Lack of Involvement of Medical Writers and the Pharmaceutical Industry in Publications Retracted for Misconduct: A Systematic, Controlled, Retrospective Study," *Current Medical Research and Opinion* 27 (2011): 1175–1182.

52. J. Scannell et al., "Diagnosing the Decline in Pharmaceutical R&D Efficiency," *Nature Review Drug Discovery* 11 (2012): 191–200; J. Scannell, "Four Reasons Drugs Are Expensive, of Which Two Are False," Innogen Working Paper no. 114, http://www.innogen.ac.uk/working-papers%5D.

53. J. DiMasi, H. Grabowski, and R. Hansen, "Innovation in the Pharmaceutical Industry: New Estimates of R&D Costs," *Journal of Health Economics* 47 (2016): 20–33.

54. J. Stiglitz, "Don't Trade Away Our Health," *New York Times*, January 30, 2015.

55. T. Stossel, "Prescription Drug Pricing: Scam or Scapegoat?" American Enterprise Institute Report, 2016, http://www.aei.org/publication/prescription-drug-pricing-scam-or-scapegoat/.

56. R. Bazell, *HER-2: The Making of Herceptin, a Revolutionary Treatment for Breast Cancer* (New York: Random House, 1998); J. LaMattina, *Drug Truths: Dispelling the Myths about Pharma R&D* (Hoboken, NJ: Wiley, 2009).

57. F. Roberts, *The Cost of Health* (London: Turnstile, 1952).

58. A. Frakt, "We Can't Have It All: The Economic Limits of Pharmaceutical Innovation," *Journal of the American Medical Association* 315 (2016): 1936–1937.

59. K. Murphy and R. Topel, "Social Value and the Speed of Innovation," *American Economic Review* 92 (2006): 433–437; K. Murphy and R. Topel, "The Value of Health and Longevity," *Journal of Political Economy* 114 (2006): 871–904.

Chapter 7

1. E. J. Calabrese and L. A. Baldwin, "The Frequency of U-shaped Dose-Responses in the Toxicological Literature," *Toxicological Sciences* 62 (2001): 330–338; E. J. Calabrese and L. A. Baldwin, "The Hormetic Dose-Response Model Is More Common than the Threshold Model in Toxicology," *Toxicological Sciences* 71 (2003): 246–250; E. J. Calabrese et al., "Hormesis Outperforms Threshold Model in National Cancer Institute Antitumor Drug Screening Database," *Toxicological*

Sciences 94 (2006): 368–378; E. J. Calabrese et al., "Hormesis Predicts Low-Dose Responses Better than Threshold Models," *International Journal of Toxicology* 27 (2008): 369–378; E. J. Calabrese et al., "Hormesis in High-Throughput Screening of Antibacterial Compounds in *E. coli*," *Human and Experimental Toxicology* 29 (2010): 667–677.

2. A biphasic response is similar to the pharmacotherapeutic model, in which low doses are beneficial and high ones are harmful.

3. Calabrese et al., "Hormesis Predicts Low-Dose Responses Better"; E. J. Calabrese et al., "Hormesis in High-Throughput Screening"; Calabrese and Baldwin, "The Frequency of U-shaped Dose-Responses"; Calabrese and Baldwin, "The Hormetic Dose-Response Model Is More Common."

4. E. J. Calabrese and R. B. Blain "The Hormesis Database: The Occurrence of Hormetic Dose Responses in the Toxicological Literature," *Regulatory Toxicology and Pharmacology* 61 (2011): 73–81.

5. E. J. Calabrese, "Historical Blunders: How Toxicology Got the Dose-Response Relationship Half Right," *Cellular and Molecular Biology* 51 (2005): 643–654.

6. T. Kuhn, *The Structure of Scientific Revolutions* (Chicago: University of Chicago Press, 1962), p. 64.

7. H. J. Muller, "Artificial Transmutation of the Gene," *Science* 66 (1927): 84–87.

8. A. R. Olson and G. N. Lewis, "Natural Reactivity and the Origin of Species," *Nature* 121 (1928): 673 674.

9. H. J. Muller and L. M. Mott-Smith, "Evidence That Natural Radioactivity Is Inadequate to Explain the Frequency of 'Natural' Mutations," *Proceedings of the National Academy of Sciences* 16 (1930): 277–285.

10. H. J. Muller, "The Problem of Genic Modification," *Supplement-band 1 der Zeitschrift fur Induktive Abstammungs und Vererbungslehre*, Manuscript Department, Lilly Library, Indiana University, Bloomington, pp. 234–260.

11. C. P. Oliver, "The Effect of Varying the Duration of X-ray Treatment upon the Frequency of Mutation," *Science* 71 (1930): 44–46; C. P. Oliver, "An Analysis of the Effect of Varying the Duration of X-ray Treatment upon the Frequency of Mutations" (Ph.D. thesis, University of Texas, Austin, 1931); F. B. Hanson and F. Heys, "An Analysis of the Effects of the Different Rays of Radium in Producing Lethal Mutations in Drosophila," *American Naturalist* 63, no. 686 (1929): 201–213; F. B. Hanson and F. Heys, "A Possible Relation between Natural (Earth) Radiation and Gene Mutations," *Science* 71, no. 1828 (1930): 43–44.

12. J. T. Patterson, "The Effects of X-rays in Producing Mutations in the Somatic Cells of Drosophila," *Science* 68 (1928): 41–43; A. Weinstein, "The Production of Mutations and Rearrangements of Genes by X-rays," *Science* 67 (1928): 376–377; L. J. Stadler, "Some Genetic Effects of X-rays in Plants," *Journal of Heredity* 21 (1930): 3–19; L. J. Stadler, "Chromosome Number and the Mutation Rule in Avena and Triticum," *Proceedings of the National Academy of Sciences* 15 (1931): 876–881.

13. H. J. Muller, "Radiation and Genetics," *American Naturalist* 64 (1930): 220–251.

14. The committee was later named the National Committee on Radiation Protection and Management.

15. E. J. Calabrese, "Key Studies Used to Support Cancer Risk Assessment Questioned," *Environmental and Molecular Mutagenesis* 52 (2011): 595–606.

16. Calabrese, "Key Studies"; E. J. Calabrese, "Origin of the Linearity No Threshold (LNT) Dose-Response Concept," *Archives of Toxicology* 87 (2013): 1621–1633; E. J. Calabrese, "How the U.S. National Academy of Sciences Misled the World Community on Cancer Risk Assessment: New Findings Challenge Historical Foundations of the Linear Dose Response," *Archives of Toxicology* 87 (2013): 2063–2081.

17. Calabrese, "How the U.S. National Academy of Sciences Misled the World Community."

18. E. J. Calabrese, "Muller's Nobel Prize Lecture: When Ideology Prevailed over Science," *Toxicological Sciences* 126 (2012): 1–4.

19. Calabrese, "Muller's Nobel Prize Lecture."

20. Calabrese, "Key Studies."

21. D. E. Uphoff and C. Stern, "The Genetic Effects of Low Intensity in Irradiation," *Science* 109 (1949): 609–610.

22. Calabrese, "Origin of the LNT Dose-Response Concept"; Calabrese, "How the U.S. National Academy of Sciences Misled the World Community." The November 2009 "climategate" emails demonstrated similar editorial actions in climate change research. No one, however, has ever done a systematic and scholarly analysis of editorial tampering.

23. E. Caspari and C. Stern, "The Influence of Chronic Irradiation with Gamma Rays at Low Dosages on the Mutation Rate in *Drosophila melanogaster*," *Genetics* 33 (1948): 75–95.

24. H. J. Muller, "Some Present Problems in the Genetic Effects of Radiation," *Journal of Cellular and Comparative Physiology* 35 Supplement, no. S2 (1950): 9–70; H. J. Muller, "Radiation Damage to the Genetic Material," *American Science* 38 (1950): 32–59; H. J. Muller, "The Nature of the Genetic Effects Produced by Radiation," in *Radiation Biology*, Vol. 1: *High Energy Radiation*, ed. A. Hollaender (New York: McGraw-Hill, 1954), chap. 7, pp. 351–473; H. J. Muller, "The Manner of Production of Mutations by Radiation," in Hollaender, *High Energy Radiation*, chap. 8, pp. 475–626.

25. See note 24.

26. J. F. Crow, "Quarreling Geneticists and a Diplomat," *Genetics* 14 (1995): 421–426. In fact, I was fortunate to share with the 95-year-old Dr. Crow my allegations about Muller's deceptions in a series of email exchanges in the months leading up to his death in January 2012. My first email to him was during the halftime of the Super Bowl, and his answer arrived before the kickoff of the second half!

27. Another example of this is in Chapter 8 of this volume, "Radiation Poisoning," which describes how the National Academy of Sciences made sure it got the politically correct answers about uranium mining in Virginia.

28. E. J. Calabrese, "On the Origins of the Linear No-Threshold (LNT) Dogma by Means of Untruths, Artful Dodges and Blind Faith," *Environmental Research* 142 (2015): 432–442; E. J. Calabrese, "LNTgate: How Scientific Misconduct by the U.S. NAS Led to Governments Adopting LNT for Cancer Risk Assessment," *Environmental Research* 148 (2016): 535–546.

29. BEAR I, "Genetic Effects of Atomic Radiation," *Science* 124 (1956): 1157–1164.

30. E. B. Lewis, "Leukemia and Ionizing Radiation," *Science* 125 (1957): 965–972.

31. G. DuShane, "Loaded Dice," *Science* 125 (1957): 964.

32. R. R. Newell, "Radiation Hazard: Its Statutory Control and Its Influence on the Future of the Human Race," *Stanford Medical Bulletin* 15 (1957): 117–122; C. Buck, "Population Size Required for Investigating Threshold Dose in Radiation-Induced Leukemia," *Science* 130 (1959): 1357–1358; A. Grendon, "Federal Radiation Council Guides and Other Exposure Standards," *American Journal of Public Health* 55 (1965): 738–747.

33. R. D. Evans, "Quantitative Inferences concerning the Genetic Effects of Radiation on Human Beings," *Science* 109 (1949): 299–304. See note 24.

34. Letter—Demerec to Muller, May 11, 1953, American Philosophical Society (APS) Library, Philadelphia.

35. M. W. Seltzer, "The Technological Infrastructure of Science" (PhD thesis, Virginia Polytechnic Institute and State University, 2007), p. 285n.

36. Letter—Dobzhansky to Demerec, August 3, 1957, Demerec files, APS Library, Philadelphia.

37. Letter—Demerec to Dobzhansky, August 9, 1957, Demerec files, APS Library, Philadelphia.

38. E. J. Calabrese, "The Road to Linearity: Why Linearity at Low Doses Became the Basis for Carcinogen Risk Assessment," *Archives of Toxicology* 83 (2009): 203–225.

39. R. E. Albert, "Carcinogen Risk Assessment in the U.S. Environmental Protection Agency," *Critical Reviews in Toxicology* 24 (1994): 75–85.

40. W. L. Russell, L. B. Russell, and E. M. Kelly, "Radiation Dose Rate and Mutation Frequency," *Science* 128 (1958): 1546–1550.

41. Andy Coghlan, "Chemistry Nobel Shared for Discovery of How DNA Repairs Itself," *New Scientist*, October 7, 2015, https://www.newscientist.com /article/dn28295-chemistry-nobel-shared-for-discovery-of-how-dna-repairs -itself/.

Chapter 8

1. Sean Cockerham and John Murawski, "Proposed Coles Hill Uranium Mine: Buried Treasure or Hidden Threat?" *McClatchy Newspapers*, December 24, 2012, http://www.mcclatchydc.com/news/nation-world/national/article24742024 .html.

2. National Academy of Sciences, *Uranium Mining in Virginia; Scientific, Technical, Environmental, Human Health and Safety, and Regulatory Aspects of Uranium Mining and Processing in Virginia* (Washington: National Research Council, 2012).

3. Marline/Umetco, "An Evaluation of Uranium Development in Pittsylvania County, Virginia (October 15, 1983)," *Technical Memoranda*, vol. 2, supplements the 1982 report, dated September 1984.

4. Andrew Rice, "Nuclear Standoff," *New Republic*, March 11, 2010.

5. Personal communication, Walter Coles Jr. to Patrick Michaels, May 4, 2012.

6. National Research Council, National Academy of Sciences, "Uranium Mining in Virginia Scientific, Technical, Environmental, Human Health and Safety, and Regulatory Aspects of Uranium Mining and Processing in Virginia," Summary Chapter, 2012, p. 9.

7. Ironically, *UMV* cites the author of this chapter for the characterization of Virginia's weather events as "extreme." The reference is given as "Hayden and Michaels, 2001." This document was an unrefereed "University of Virginia News Letter" written in 1981 called "Virginia's Climate." The only change in 2001 was to update the 30-year climate "normals" to 1971–2000.

8. National Research Council, National Academy of Sciences, "Uranium Mining in Virginia," p. 41.

9. National Weather Service, "Historical Hurricane Tracks, 1933–1998, Virginia and the Carolinas," Document available from the National Weather Service Forecast office in Wakefield, Virginia.

10. Historical maps of North Atlantic tropical cyclones are available at http://weather.unisys.com/hurricane/atlantic/index.php.

11. F. K. Schwartz, "The Unprecedented Rains in Virginia Associated with the Remnants of Hurricane Camille," *Monthly Weather Review* 98 (1970): 851–859.

12. L. S. Eaton et al., "Role of Debris Flow in Long-Term Landscape Denudation in the Central Appalachians of Virginia," *Geology* 31 (2003): 339–342.

13. National Oceanic and Atmospheric Administration, "Historical Hurricane Tracks," 2013, http://www.csc.noaa.gov/hurricanes/#.

14. I have covered other excruciating details and mistaken interpretations of the potential for flooding in a comprehensive peer-reviewed paper. See P. J. Michaels et al., "Misleading Geophysical Arguments from Uranium Mining in Virginia," *Journal of Research in Environmental and Earth Sciences* 1 (2014): 12–24.

15. United States Geological Survey, "Virginia Earthquake Information," 2012, http://earthquake.usgs.gov/earthquakes/states/?region=Virginia.

Chapter 9

1. Richard Schwartz, Attorney for Northern Dynasty Minerals Ltd., Request for Investigation Concerning EPA Bristol Bay Watershed Assessment, to the Inspector General, Environmental Protection Agency, January 9 and February 18, 2014, http://www.northerndynastyminerals.com/site/assets/files/4568/plp_response_to_final_bbwa_april2014.pdf.

2. Andrew Watson, "Mining Properties and Prospects," *Geology for Investors*, February 2016.

3. Daniel McGroarty, "Miners Struggle with a Federal Cave-In," *Wall Street Journal*, July 24, 2014.

4. McGroarty, "Miners Struggle."

5. Daniel McGroarty, "EPA's Bristol Bay Watershed Assessment: A Factual Review of a Hypothetical Scenario," Testimony presented to the U.S. House of Representatives Committee on Science, Space and Technology Subcommittee, August 1, 2013, http://americanresources.org/epas-bristol-bay-watershed-assessment-a-factual-review-of-a-hypothetical-scenario/.

6. Environmental Protection Agency, Clean Water Act, Section 404(c), "Veto Authority," https://www.epa.gov/sites/production/files/2016-03/documents/404c.pdf.

7. Doug Steding, "EPA's Initiation of a Clean Water Act Section 404(c) Review for the Mining of the Pebble Deposit: What Is the History of EPA's Other 404(c) Determinations?" *Science, Law, and the Environment*, March 3, 2014, http://www.sciencelawenvironment.com/2014/03/epas-initiation-of-a-clean-water-act-section-404c-review-for-the-mining-of-the-pebble-deposit-what-is-the-history-of-epas-other-404c.

8. Steding, "EPA's Initiation of a Clean Water Act Review."

9. Kimberley Strassel, "The EPA's Pebble Blame Game," *Wall Street Journal*, May 21, 2015.

10. Alex DeMarban, "Pebble Unveils Long-Awaited Smaller Mine Plan," *Anchorage Daily News*, October 4, 2017.

11. Ned Mamula and Patrick J. Michaels, "Special Report; A Green Mess: Is EPA in Hot Water over Alaska's Bristol Bay?" *American Spectator*, February 11, 2016.

12. Pebble Partnership, Environmental Baseline Document, https://pebbleresearch.com/ and https://pebbleresearch.com/download/.

13. United States District Court for the District of Alaska, videotaped deposition of Philip North, https://www.peer.org/assets/docs/ak/4_4_16_Day1_Pebble_depo.pdf.

14. U.S. House of Representatives, Committee on Science, Space and Technology, deposition of Philip A. North, April 15, 2016, https://science.house.gov/sites/republicans.science.house.gov/files/documents/Deposition%20Transcript.pdf; Valerie Richardson, "EPA Accused of Collusion after Staffer Admits He Aided Pebble Mine Foes," *Washington Times*, April 28, 2016.

15. U.S. House of Representatives, Committee on Science, Space and Technology, deposition of Dennis McLerran, April 28, 2016, http://docs.house.gov/meetings/SY/SY00/20160428/104889/HHRG-114-SY00-20160428-SD002.pdf; Kimberley Strassel, "The EPA's Own Email Problem," *Wall Street Journal*, August 27, 2015; Tim Sohn, "The EPA Ecologist Who Became a Wanted Man," *New Yorker*, May 3, 2016.

16. Schwartz, Request for Investigation Concerning EPA Bristol Bay Watershed Assessment.

17. U.S. Environmental Protection Agency, "National Environmental Policy Act," https://www.epa.gov/nepa.

18. U.S. Environmental Protection Agency, "National Environmental Policy Act—Policies and Guidance," https://www.epa.gov/nepa/national-environmental-policy-act-policies-and-guidance.

19. USGS reviewer comments are not attributed to any one person publicly.

20. U.S. Environmental Protection Agency, Office of Inspector General, "EPA's Bristol Bay Watershed Assessment: Obtainable Records from the Office of the Inspector General Show EPA Followed Required Procedures without Bias or Predetermination, but a Possible Misuse of Position Noted," January 13, 2016, https://www.epa.gov/sites/production/files/2016-01/documents/20160113-16-p-0082.pdf.

21. U.S. EPA, "EPA's Bristol Bay Watershed Assessment."

22. U.S. House of Representatives, Committee on Oversight and Government Reform, November 4, 2015, https://oversight.house.gov/wp-content/uploads/2015/11/2015-11-04-JC-CL-JJ-to-McCarthy-EPA-Bristol-Bay-due-11-18.pdf.

23. U.S. EPA, "EPA's Bristol Bay Watershed Assessment."

24. U.S. EPA.

25. Julian Hattem, "Former EPA Chief under Fire for New Batch of 'Richard Windsor' Emails," *The Hill*, May 1, 2013.

26. Pebble Limited Partnership, "Examining the Facts: A Response to the U.S. Environmental Protection Agency's (EPA) Office of Inspector General Report Regarding EPA's Bristol Bay Watershed Assessment," letter dated February 29, 2016, https://science.house.gov/sites/republicans.science.house.gov/files/documents/02.29.16%20SST%20Letter%20to%20EPA%20OIG%20Elkins%20re%20Bristol%20Bay%20Rep.

27. U.S. EPA, "EPA's Bristol Bay Watershed Assessment."

28. Pebble Limited Partnership, "Examining the Facts."

29. U.S. House of Representatives, Committee on Science, Space and Technology, letter to EPA administrator requesting withdrawal of Pebble Project veto, February 22, 2017, https://science.house.gov/sites/republicans.science.house.gov/files/documents/02.29.16%20SST%20Letter%20to%20EPA%20OIG%20Elkins%20re%20Bristol%20Bay%20Rep.

30. Kalhan Rosenblatt, "Trump Signs Executive Order to Begin Water Rule Rollback," NBC News, February 28, 2017, http://www.nbcnews.com/news/us-news/trump-signs-executive-order-begin-water-rule-rollback-n726781.

31. U.S. EPA, "2017 Settlement Agreement between EPA and Pebble Limited Partnership," https://www.epa.gov/bristolbay/2017-settlement-agreement-between-epa-and-pebble-limited-partnership.

Chapter 10

1. This includes $2.5 billion for the U.S. Global Change Research Program, plus additional money that only seems justified by the climate issue, such as funding the Global Environmental Facility and the Clean Technology Fund. When these are included, total spending on climate issues in 2013 was $6.4 billion. The entire annual total given for all categories in the Report to Congress is $22.6 billion. Office of Management and Budget, "Federal Climate Change Expenditures," Report to Congress, https://obamawhitehouse.archives.gov/sites/default/files/omb/assets/legislative_reports/fcce-report-to-congress.pdf.

2. *Massachusetts v. Environmental Protection Agency*, 549 U.S. 497 (2007).

3. NEEM Community Members, "Eemian Interglacial Reconstructed from a Greenland Folded Ice Core," *Nature* 493 (2013): 489–494.

4. A. Schmittner et al., "Climate Sensitivity Estimated from Temperature Reconstructions of the Last Glacial Maximum," *Science* 334 (2011): 1385–1388; J. C. Hargreaves et al., "Can the Last Glacial Maximum Constrain Climate Sensitivity?" *Geophysical Research Letters* 39 (2012): L24702.

5. U.S. Environmental Protection Agency, "Technical Support Document for the Endangerment and Cause or Contribute Findings for Greenhouse Gases," 2009, https://www.epa.gov/ghgemissions/technical-support-document-endangerment-and-cause-or-contribute-findings-greenhouse.

6. J. R. Christy, "Testimony before the House Committee on Science, Space and Technology, 2017," March 29, 2017; J. R. Christy, "State of the Climate in 2017,"

Special supplement to *Bulletin of the American Meteorological Society* 98 (2017): Figure S10.

7. T. F. Stocker et al., *Climate Change 2013: The Physical Science Basis. Contribution of Working Group I to the Fifth Assessment Report of the Intergovernmental Panel on Climate Change* (Cambridge: Cambridge University Press, 2013).

8. Ron Clutz, "Temperatures according to Climate Models," *Science Matters* (blog), March 24, 2015, https://rclutz.wordpress.com/2015/03/24/temperatures-according-to-climate-models/; T. Andrews et al., "Forcing, Feedbacks and Climate Sensitivity in a New Generation of Climate Models," *Geophysical Research Letters* 39, no. 9 (May 2012): https//doi.org/10.1029/2012GL051607.

9. A note on units. The entire weight of the atmosphere is approximately 1,000 hectopascals (hPa). Therefore when the balloon senses 500 hPa, about half of the atmosphere (by weight) is above it, and half is below. Sea level is approximately 1,000 hPa; the second level (not shown on the y axis) is 850 hPa (around 5,000 ft.), followed by 700 hPa (10,000 ft.), 500 hPa (18,000 ft.), 250 hPa (35,000 ft.), and 100 hPa (52,000 ft.).

10. Z. Zhu et al., "Greening of the Earth and Its Drivers," *Nature Climate Change* 6 (2016): 791–795. This is the same phenomenon that greenhouse owners take advantage of: many greenhouses are run with three times the ambient ("natural") concentration of carbon dioxide in order to stimulate plant growth.

11. S. Munier et al., "Satellite Leaf Area Index: Global Scale Analysis of the Tendencies per Vegetation Type over the Last 17 Years," *Remote Sensing* 424 (2018), https://doi.org/10.3390/rs10030424. See also Patrick J. Michaels, "The Most Amazing Greening on Earth," *Climate Etc.* (blog), September 19, 2018, https://judithcurry.com/2018/09/19/the-most-amazing-greening-on-earth/.

12. J. T. Houghton et al., eds., *Climate Change 1995: The Science of Climate Change. Contribution of Working Group I to the Second Assessment Report of the Intergovernmental Panel on Climate Change* (Cambridge: Cambridge University Press, 1996).

13. P. Voosen, "Climate Scientists Open Up Their Black Boxes to Scrutiny," *Science* 354 (2016): 401–402.

14. Voosen, "Climate Scientists Open Up," p. 401.

15. T. Mauritsen et al., "Tuning the Climate of a Global Model," *Journal of Advances in Modeling Earth Systems* 4 (2012), https://doi.org/10.1029/2012MS000151.

16. F. Hourdin et al., "The Art and Science of Climate Model Tuning," *Bulletin of the American Meteorological Society*, March 31, 2017, https://doi.org/10.1175/BAMS-D-15-00135.1.

17. From Voosen, "Climate Scientists Open Up," p. 402: "Indeed, whether climate scientists like to admit it or not, nearly every model has been calibrated precisely to 20th century climate records—otherwise it would have ended up in the trash. 'It is fair to say that all models have tuned it,' says Isaac Held, a scientist at the Geophysical Fluid Dynamics Laboratory, another prominent modeling center in Princeton, New Jersey."

18. F. Hourdin et al., "The Art and Science of Climate Model Tuning."

19. Judith Curry, "The Art and Science of Climate Model Tuning," *Climate Etc.* (blog), August 1, 2016, https://judithcurry.com/2016/08/01/the-art-and-science-of-climate-model-tuning/.

20. All of this is documented in P. J. Michaels, "Science or Political Science? An Assessment of the U.S. National Assessment of the Potential Consequences of Climate Variability and Change," in *Politicizing Science: The Alchemy of Policymaking*, ed. M. Gough (Stanford, CA: Hoover Institution Press, 2003).

21. Thomas R. Karl et al., *Global Climate Change Impacts in the United States: A State of the Knowledge Report from the U.S. Global Change Research Program* (New York: Cambridge University Press, 2009).

22. Patrick J. Michaels, ed., *Addendum: Global Climate Impacts in the United States* (Washington: Cato Institute, Center for the Study of Science, 2012), pp. 139–145, 158–161.

23. P. J. Michaels and P. C. Knappenberger, *Lukewarming: The New Climate Science That Changes Everything* (Washington: Cato Institute, 2016).

24. T. Karl et al., *Global Climate Change Impacts in the United States*.

25. It is by far the most frequently cited document in the Technical Support Document.

26. J. J. Melillo et al., eds., *Climate Change Impacts in the United States: The Third National Climate Assessment* (Washington: Government Printing Office, 2014), introduction.

27. A fourth National Assessment was in press at the time of this writing. It contains the same problems as the second and third National Assessments.

Chapter 11

1. Brady Dennis and Juliet Eilperin, "Scott Pruitt Blocks Scientists with EPA Funding from Serving as Agency Science Advisors," *Washington Post*, October 31, 2017.

2. Office of Management and Budget, "2015 Draft Report to Congress on the Benefits and Costs of Federal Regulations and Agency Compliance with the Unfunded Mandates Reform Act," p. 13.

3. Susan E. Dudley, "OMB's Reported Benefits of Regulation: Too Good to Be True?" *Regulation* 36, no. 2 (2013): 26–30. As Dudley explains (p. 29), the private benefits from regulation included in Figure 11.2 are private savings generated by public regulations. An important example of such private benefits is the fuel savings consumers realize from tougher automobile mileage standards. As can be seen in Figure 11.2, such private benefits were essentially nonexistent in OMB cost-benefit analysis until the Obama administration. Dudley provides an explanation and critique of the methodological change that led to this increase in private benefits.

4. U.S. EPA, *Integrated Science Assessment (ISA) for Particulate Matter, Final Report*, December 2009, https://cfpub.epa.gov/ncea/risk/recordisplay.cfm?deid =216546.

5. Clean Air Act, 42 U.S.C. §7408(b)(1).

6. Clean Air Act, §7408(a)(2).

7. Pub. L. 95–95, title I, § 106, August 7, 1977, 91 Stat. 691.

8. Clean Air Act, §7409(d)(1).

9. Criteria pollutants are those for which there is a defined quantitative maximum emission.

10. J. E. Goodman et al., "Evaluation of the Causal Framework Used for Setting National Ambient Air Quality Standards," *Critical Reviews in Toxicology* 43 (2013): 830.

11. Goodman et al., "Evaluation of the Causal Framework," p. 831.

12. U.S. EPA, *Integrated Science Assessment (ISA) for Ozone and Related Photochemical Oxidants (Final)*, EPA/600/R–10/076F, February 2013, Table II.

13. Goodman et al., "Evaluation of the Causal Framework," p. 834.

14. Susan E. Dudley and Marcus Peacock, "Regulatory Science and Policy: A Case Study of the National Ambient Air Quality Standards," *Supreme Court Economic Review* 24 (2018): 49–99.

15. Goodman et al. "Evaluation of the Causal Framework," p. 834.

16. Its seven members must include at least one member of the National Academy of Sciences, one physician, and one person representing state air pollution control agencies. Clean Air Act, §7409(d)(2)(A).

17. U.S. EPA, *Integrated Science Assessment (ISA) for Particulate Matter, Final Report*, pp. 1–14.

18. EPA, "Regulatory Impact Analysis for the Final Revisions to the National Ambient Air Quality Standards for Particulate Matter" (December 2012, replaced on February 28, 2013, to add Appendix 3.A, which was inadvertently left out and to correct the document number to EPA-452/R-12-005), Section 5, https://www3.epa.gov/ttnecas1/regdata/RIAs/finalria.pdf.

19. The two studies are Daniel Krewski et al., "Extended Follow-up and Spatial Analysis of the American Cancer Society Study Linking Particulate Air Pollution and Mortality," Health Effects Institute Research Report no. 140, Boston, 2009; and J. Lepuele et al., "Chronic Exposure to Fine Particles and Mortality: An Extended Follow-up of the Harvard Six Cities Study from 1974 to 2009," *Environmental Health Perspectives* 120 (2012): 965–970.

20. EPA, *Integrated Science Assessment (ISA) for Particulate Matter*, pp. 6–182.

21. Meredith Franklin, Ariana Zeka, and Joel Schwartz, "Association between $PM_{2.5}$ and All-Cause and Specific-Cause Mortality in 27 U.S. Communities," *Journal of Exposure Science and Environmental Epidemiology* 17 (2007): 279–287.

22. To be precise, the relationship was between mortality and the $PM_{2.5}$ level one day before the death (and the control measurement date).

23. Franklin, Zeka, and Schwartz, "Association between $PM_{2.5}$ and Mortality," p. 282.

24. EPA, *Integrated Science Assessment (ISA) for Particulate Matter*, pp. 6–175.

25. As for the significance of air conditioning to the observed mortality relationship, it is true that with 90.8 percent of homes air conditioned, Phoenix had the second-highest air conditioning rate among all 27 cities. At about 50 percent, Milwaukee was in the bottom third of air-conditioning prevalence. Although Phoenix and Milwaukee have the largest estimated increases in mortality due to fine PM, they are at opposite ends of the spectrum in air-conditioning prevalence.

26. See Social Security Administration, Actuarial Life Table, 2014, www.ssa.gov/oact/STATS/table4c6.html.

27. Bart Ostro et al., "Fine Particulate Air Pollution and Mortality in Nine California Counties: Results from CALFINE," *Environmental Health Perspectives* 114 (2006): 29–33.

28. Lisa K. Baxter et al., "Influence of Exposure Differences on City-to-City Heterogeneity in PM$_{2.5}$-Mortality Associations in U.S. Cities," *Environmental Health* 16 (2017): 1.

29. Krewski et al., "Extended Follow-up and Spatial Analysis"; Lepuele et al., "Chronic Exposure to Fine Particles."

30. Lepuele et al., "Chronic Exposure to Fine Particles."

31. As explained in Krewski et al., "Extended Follow-up and Spatial Analysis," p. 49, the marginal change of 1.5 μg/m^3 was used because this is the difference between the 90th and 10th percentiles in the full 28-county, three-year average PM$_{2.5}$ levels. Average relative risk is 1.0. A value above 1.0 indicates a greater risk from a certain factor (like fine PM); 2.0 means an individual has twice the risk. For values below 1.0, the risk is lower from a certain factor.

32. Mortality data due to diabetes and endocrine-related deaths show a very large spread, with the vast majority of the statistical range in the positive area. Although in the figure PM$_{2.5}$ has a relatively large (>1) impact in increasing endocrine and diabetes-related mortality, Krewski et al., "Extended Follow-up and Spatial Analysis," p. 49, point out that these effects are "confounded" by the ecologic covariates (the neighborhood socioeconomic status proxies). What they mean by this can be seen from their Table 15 on p. 50: when all seven of their ecologic covariates are included, the 95 percent confidence interval for diabetes, for example, includes a 27 percent decline in relative risk due to fine particulate exposure.

33. For example, zip code 90049 includes the ultra-wealthy neighborhoods of Brentwood, Mandeville Canyon, Westwood, and Bel Air, and also much less wealthy neighborhoods such as South Valley and Mid-City. That zip code's median income ranged from about $208,000 in Bel Air, which is 80 percent white, to $44,000 in Mid-City, which is about 10 percent white. L.A. Biz, *Here Are Los Angeles' Wealthiest Zip Codes* (Interactive Map), available at https://www.bizjournals.com/losangeles/news/2017/02/09/here-are-los-angeles-wealthiest-zip-codes-i.html.

34. Krewski et al., "Extended Follow-up and Spatial Analysis," pp. 6–63.

35. For example, a single Brooklyn neighborhood, Borough Park, includes the zip codes 11204, 11218, 11219, and 11230. For full zip code definitions of New York neighborhoods, see New York State Department of Health, "Zip Code Definitions of New York Neighborhoods," https://www.health.ny.gov/statistics/cancer/registry/appendix/neighborhoods.htm.

36. Krewski et al., "Extended Follow-up and Spatial Analysis," p. 54, Table 17.

37. This is now known as "p-hacking," which means switching and transforming variables as well as using different model variates until some statistically "significant" (p < .05) relationship falls out of the data by chance. On average, from a field of 20 variables composed of random numbers, the 5 percent confidence level will be reached on half of the analyses, as 1 in 20 = .05.

38. As a leading textbook puts it, "By increasing sample size, smaller and smaller effects will be found to be statistically significant, until at very large samples sizes almost any effect is significant." J. Hair et al., *Multivariate Data Analysis*, 6th ed. (New York: Pearson, 2006), p. 11.

39. Ostro et al., "Fine Particulate Air Pollution and Mortality," p. 32.

40. To see this problem—which may be summarized as too coarse a level of analysis—consider subsequent work that the EPA also relied on prominently

in its fine PM risk assessment: Meredith Franklin et al., "The Role of Particle Composition on the Association between PM$_{2.5}$ and Mortality," *Epidemiology* 19 (2008): 680–689. Franklin et al. applied the same methods of statistical analysis to a slightly different set of cities. In some of their regression analyses, they controlled for what they called "community-specific" parameters such as median household income and percentage of the population above 65 years of age. However, they define a "community" as an entire county, as they did in an earlier paper (Franklin et al., "Association between PM$_{2.5}$ and Mortality"). *Thus they consider Los Angeles County—which includes 88 cities within its borders ranging in socioeconomic status from Beverly Hills and Santa Monica to Compton and Inglewood—to be one "community."* The SES data here are so coarse as to be useless.

41. It seems odd that Krewski et al., "Extended Follow-up and Spatial Analysis," p. 30, found that having less than a grade 12 education in their dataset was negatively correlated with exposure to fine PM, whereas the percentage of the population that was nonwhite and median household income were both positively correlated with fine PM exposure levels. As for their 44 individual covariates—individual variables such as smoking history, diet, and alcohol consumption—the vast majority of these can and do change over a person's lifespan, and yet they used data that apparently were recorded once, in a 1982 survey. This is far too coarse a measure of lifestyle behaviors that are crucial to disease and mortality risk.

42. Iyad Kheirbek et al., "The Contribution of Motor Vehicle Emissions to Ambient Fine Particulate Matter Public Health Impacts in New York City: A Health Burden Assessment," *Environmental Health* 15 (2016): 89.

43. J. Schwartz, "Air Pollution and Daily Mortality in Birmingham, Alabama," *American Journal of Epidemiology* 13 (1993): 1136–1147.

44. R. L. Smith et al., "Regression Models for Air Pollution and Daily Mortality: Analysis of Data from Birmingham, Alabama," *Environmetrics* 11 (2000): 719–743.

45. S. Stanley Young, "Air Quality Environmental Epidemiology Studies Are Unreliable," *Regulatory Toxicology and Pharmacology* 86 (2017): 178.

46. Louis Anthony Cox and Douglas A. Popken, "Has Reducing Fine Particulate Matter and Ozone Reduced Mortality Rates in the United States?" *Annals of Epidemiology* 25 (2015): 162–173; and Tony Cox, Douglas Popken, and Paolo Ricci, "Temperature, Not Fine Particulate Matter (PM$_{2.5}$), Is Causally Associated with Short-Term Acute Daily Mortality Rates: Results from One Hundred United States Cities," *Dose-Response* 11 (2013): 319–343.

47. For these trends in all counties in the 15 largest U.S. states, see Cox and Popken, "Has Reducing Fine Particulate Matter Reduced Mortality Rates?" p. 166.

48. Goran Krstic, "A Reanalysis of Fine Particulate Matter Air Pollution versus Life Expectancy in the United States," *Journal of the Air and Waste Management Association* 63 (2013): 133–135.

49. S. Stanley Young and Jessie Q. Xi, "Assessing Geographic Heterogeneity and Variable Importance in an Air Pollution Data Set," *Statistical Analysis and Data Mining* 6 (2013): 375–386.

50. S. Stanley Young, R. L. Obenchain, and C. G. Lambert, "Bias and Response Heterogeneity in an Air Quality Dataset," https://www.researchgate.net/profile/S_Young.

51. Cox, Popken, and Ricci, "Temperature, Not Fine Particulate Matter," p. 334.

52. Borek Puza et al., "Constrained Confidence Intervals in Time Series Studies of Mortality and Air Pollution," *Environment International* 37 (2011): 204.

53. See also E. J. Calabrese and R. B. Blain. "The Hormesis Database: The Occurrence of Hermetic Dose Responses in the Toxicological Literature," *Regulatory Toxicology and Pharmacology* 61 (2011): 73–81.

54. Sunlight is a prime example, as it promotes health in small quantities by completing the synthesis of vitamin D, but in large quantities can be fatal in both the short-term (dehydration) and long-term (carcinogenesis) exposure windows. It is obvious that in many cases evolution has worked to wring benefits out of small exposures to things that are ubiquitous in our environment. Given that fine PM is also ubiquitous, it may be another "hormetic" species.

55. Louis Anthony Cox Jr., "Hormesis for Fine Particulate Matter (PM 2.5)," *Dose Response* 10 (2012): 209–218.

56. Cox, "Hormesis for Fine Particulate Matter," pp. 212–213. The data here, like those for fine PM, are too noisy to be dispositive of hormesis; see Figure 11.6.

57. H. Li et al., "PM 2.5 and PM 10 Emissions from Soils by Wind Erosion," *Aeolian Research* 19(B) (2015): 171–182.

58. Suresh H. Moolgavkar, "A Review and Critique of the EPA's Rationale for a Fine Particle Standard," *Regulatory Toxicology and Pharmacology* 42 (2005): 123–144.

59. Moolgavkar, "A Review and Critique," p. 139; G. Koop and Lise Tole, "An Investigation of Thresholds in Air Pollution—Mortality Effects," *Environmental Modelling and Software* 21 (2006): 1662–1673.

60. Moolgavkar, "A Review and Critique," p. 139.

61. Moolgavkar, p. 139.

62. D. Thomas et al., "Bayesian Model Averaging in Time-Series Studies of Air Pollution and Mortality," *Journal of Toxicology and Environmental Health* 70 (2007): 311–315; Moolgavkar, "A Review and Critique," p. 140.

63. Moolgavkar, "A Review and Critique," p. 141.

64. Pope, Ezzati, and Dockery, "Fine-Particulate Air Pollution and Life Expectancy," p. 377.

65. National Research Council, *Research Priorities for Airborne Particulate Matter I. Immediate Priorities and a Long-Range Research Portfolio* (Washington: National Academies Press, 1998).

66. U.S. EPA, "Research Center for Particulate Air Pollution and Health," EPA Grant Number: R827355, Airborne PM–Northwest Research Center for Particulate Air Pollution and Health, https://cfpub.epa.gov/ncer_abstracts/index .cfm/fuseaction/display.highlight/abstract/1088.

67. U.S. EPA, "NYU–EPA PM Center: Health Risks of PM Components," EPA Grant Number: R827351, EPA NYU PM Center: Health Risks of PM Componenets, https://cfpub.epa.gov/ncer_abstracts/index.cfm/fuseaction/display.abstractDetail /abstract/1089.

68. U.S. EPA, "Ultrafine Particles: Characterization, Health Effects and Pathophysiological Mechanisms," EPA Grant Number: R827354, Airborne PM–Rochester PM Center, https://cfpub.epa.gov/ncer_abstracts/index.cfm /fuseaction/display.abstractDetail/abstract/1098/report/0.

69. U.S. EPA, "Southern California Particle Center and Supersite (SCPCS)," EPA Grant Number: R827352, Southern California Particle Center and Supersite, https://cfpub.epa.gov/ncer_abstracts/index.cfm/fuseaction/display.highlight /abstract/1087.

70. U.S. EPA, "Ambient Particle Health Effects: Exposure, Susceptibility, and Mechanisms," EPA Grant Number: R827353, EPA Harvard Center for Ambient Particle Health Effects, https://cfpub.epa.gov/ncer_abstracts/index.cfm/fuseaction /display.abstractDetail/abstract/1088.

71. National Research Council, Committee on Research Priorities for Airborne Particulate Matter, "Research Priorities for Airborne Particulate Matter: I. Immediate Priorities and a Long-Range Research Portfolio iii," supra.

72. U.S. EPA, "Harvard Particle Center," EPA Grant Number: R832416, Harvard Particle Center, https://cfpub.epa.gov/ncer_abstracts/index.cfm/fuseaction /display.abstractDetail/abstract7739.

73. U.S. EPA, "Rochester PM Center: Source-Specific Health Effects of Ultrafine/Fine Particles," EPA Grant Number: R832415, Center: Rochester PM Center, https://cfpub.epa.gov/ncer_abstracts/index.cfm/fuseaction/display.abstractDetail /abstract/7758.

74. U.S. EPA, "Southern California Particle Center (SCPC)," EPA Grant Number: R832413, Southern California Particle Center, https://cfpub.epa.gov /ncer_abstracts/index.cfm/fuseaction/display.abstractDetail/abstract/7740.

75. U.S. EPA, "Johns Hopkins Particulate Matter Research Center," EPA Grant Number: R832417, Johns Hopkins Particulate Matter Research Center, https://cfpub.epa.gov/ncer_abstracts/index.cfm/fuseaction/display.abstractDetail /abstract/7738.

76. U.S. EPA, "San Joaquin Valley Aerosol Health Effects Research Center (SAHERC)," EPA Grant Number: R832414, San Joaquin Valley Aerosol Health Effects Research Center (SAHERC), https://cfpub.epa.gov/ncer_abstracts/index .cfm/fuseaction/display.abstractDetail/abstract/7741.

77. For a list of the grants, see U.S. EPA, Health Effects Institute, https:// cfpub.epa.gov/ncer_abstracts/index.cfm/fuseaction/recipients.display/location _id/3337. According to HEI, its private funding, primarily from automobile companies roughly matches the amount contributed by the EPA and other government sponsors; see the Health Effects Institute website, https://www.healtheffects .org/about/sponsors.

78. University of Washington, "Multi-Ethnic Study of Atherosclerosis and Air Pollution," MESA Air Home, https://deohs.washington.edu/mesaair/home.

79. Franklin, Zeka, and Schwartz, "Association between $PM_{2.5}$ and Mortality"; Franklin et al., "The Role of Particle Composition"; Antonella Zanobetti and Joel Schwartz, "The Effect of Fine and Coarse Particulate Air Pollution on Mortality: A National Analysis," *Environmental Health Perspectives* 117 (2009): 897–903.

80. Pope, Ezzati, and Dockery, "Fine-Particulate Air Pollution and Life Expectancy," supported in part by grants from the Association of Schools of Public Health (U36/CCU300430-23), the Harvard Environmental Protection Agency (EPA) Particulate Matter Center (EPA RD832416), and the National Institute of Environmental Health Sciences (ES0002); and funds from the Mary Lou Fulton

Professorship, Brigham Young University; C. A. Pope III et al., "Cardiovascular Mortality and Long-Term Exposure to Air Pollution: Epidemiological Evidence of General Pathophysiological Pathways of Disease," *Circulation* 109 (2004): 71–77, supported in part by grants from the National Institute of Environmental Health Sciences (NIH; ES09560, ES00260, and ES08129), the U.S. Environmental Protection Agency (R827351), and a Health Effects Institute contract; C. A. Pope III et al., "Particulate Air Pollution as a Predictor of Mortality in a Prospective Study of U.S. Adults," *American Journal of Respiratory and Critical Care Medicine* 151 (1995): 669–674, supported in part by National Institute of Environmental Health Sciences Grants ES-OOOO2 and ES-01108 and by Environmental Protection Agency Cooperative Agreements CR-811650 and CR-818090; F. Laden et al., "Reduction in Fine Particulate Air Pollution and Mortality: Extended Follow-up of the Harvard Six Cities Study," *American Journal of Respiratory and Critical Care Medicine* 173 (2006): 667–672, supported by grants from the U.S. Environmental Protection Agency (EPA; R827353) and the National Institute of Environmental Health Sciences (ES00002).

81. Morton Lippman of the NYU Center suggested two studies, including C. Arden Pope et al., "Cardiovascular Mortality and Long Term Exposure to Particulate Air Pollution: Epidemiological Evidence of General Pathophysiological Pathways of Disease," *Circulation* 109 (2004): 71–77; Speizer of the Harvard Center suggested four new studies, including Laden et al., "Reduction in Fine Particulate Air Pollution"; and Krewski et al., "Mortality and Longterm Exposure to Ambient Air Pollution: Ongoing Analysis on the American Cancer Society Cohort," *Journal of Toxicology and Environmental Health, Part A* 68 (2005): 1093–1109. Ron White suggested F. Dominici et al., "Fine Particulate Air Pollution and Hospital Admissions for Cardiovascular and Respiratory Diseases," *Journal of the American Medical Association* 295 (2006): 1127–1134.

82. McClellan brought another study, by J. E. Enstrom, "Fine Particulate Air Pollution and Total Mortality among Elderly Californians, 1973–2002," *Inhalation Toxicology* 17 (2005): 803–816, to the attention of the committee, noting it "does not support a current relationship between fine PM and total mortality in elderly Californians, but maybe a small effect before 1983."

83. For example, the Southern California Particle Center and Supersite's 1998 website stated that "the overall objective of the Southern California Particle Center and Supersite (SCPCS) is to bring together outstanding scientists from the leading universities in Southern California to identify and conduct high priority research to better understand the effects of particulate matter (PM) and ensure protection of public health. . . . By improving our fundamental and observational understanding of the complex relation between particle exposure and human health, it is our goal to lay a firm scientific foundation for effective intervention strategies for public health protection." Other centers have similar objectives.

Note: Information in figures and tables is indicated by f and t; n designates a numbered note.

American Cancer Society (ACS)
study, 270, 284
American Heart Association (AHA)
backing away from salt-hypertension
hypothesis, 105
on reduction of salt intake, 96, 99
scientific consensus on health and
diet and, 55, 56, 57
*American Journal of Clinical
Nutrition*, 88
American Medical Association, 48
on anti-salt propaganda, 96
on government recommendations
on salt intake, 88
American Meteorological Society, 250
American Pain Society, 131
American Philosophical Society,
292n18
American Public Health Association
(APHA), 78
American Society for Clinical
Investigation, 163
American X-ray and Radium
Protection Committee, 191–92
Andrus, E. Cowles, 87
angiotensin, 114–15
Annals of Internal Medicine, 97, 130
anti-nuclear environmentalists, 210
antioxidants, exposure to fine PM and
increased production of, 279
anti-salt movement. *See also* salt-
hypertension hypothesis
Dahl and, 70–74
"devil shift" in rhetoric and,
99–103
refusal to consider research
refuting conclusions, 84–85
anti-salt propaganda, 95–97
APHA. *See* American Public Health
Association (APHA)
Appel, Lawrence J., 103, 115–16
Arrowsmith (Lewis), 163
arsenicals, 162
"The Art and Science of Model
Tuning" (Hourdin), 249
Ashbaugh, Lowell, 289t
aspirin, 162

"An Assessment of the Potential
Mining Impacts on Salmon
Ecosystems of Bristol Bay,
Alaska" (EPA), 225
Association of American Medical
Colleges, 177
Association of American Physicians,
163
atherosclerosis
heart disease and, 40–41
Keys's research and first model for,
39–45
proposed causes of, 42–44, 44f
"Atherosclerosis: A Problem in Newer
Public Health" (Keys), 39–41
atherosclerotic heart disease, death
rates per 100,000 people in
20th century, 40f
Atkins, Robert, 54, 57
Atomic Energy Act, 221
Atomic Energy Commission (AEC),
195, 205–6
ATSDR. *See* Agency for Toxic
Substances and Disease Registry
(ATSDR)
Australian open-pit mine, 259f
auto emissions of carbon dioxide,
limits on, 240
Avol, Ed, 289t

Bache, Alexander, 292n18
Bacon, Francis
Bush and, 24
on four-stage process of science, 32
on knowledge as power, 10, 23
linear model and, 13
science as public good and, 11–12
Bairoch, Paul, 18
Barker, David, 52–53
Barouch, Winnie, 103
Barr, William, 151
basic research
Bush on, 24–25, 163–64, 166, 173
downsizing, 181
intrinsically interesting phenomena
and, 174
medical innovation *vs.*, 166–67

carbohydrates
 abstention from fats and increased
 consumption of, 46, 118
 elevation of cholesterol and, 50
 heart disease and, 54
 obesity and, 56
carbon dioxide, effect on leaf area
 index, 246–47
carbon dioxide, EPA's finding of
 endangerment from, 239–56
 errors in climate models used,
 242–45, 243f, 244f
 global greening, 246–47, 246f
 history of endangerment finding,
 239–47
 model tuning and, 247–50
 systematic flaws in U.S. National
 Climate Assessments, 251–56
carbon monoxide, as criteria air
 pollutant, 262
carcinogens and chemicals, regulation
 of, 6, 185–207
 academics and self-interest science,
 203–5
 cancer risk assessment, 195–99, 205–7
 EPA Cancer Assessment Group,
 205–7
 linear no-threshold model, 186–87,
 186f
 linear no-threshold model origin
 and implications, 190–92
 Manhattan Project and linear
 no-threshold model, 192–95
 National Academy of Sciences
 BEAR I Genetic Panel, 185,
 199–203, 205–7
 risk assessment and, 185–89, 186f
cardiovascular disease. *See also* heart
 disease
 causes of differences between
 nations, 307n159
 risk identification, 168
 study of advances in treatment, 170
CASAC. *See* Clean Air Scientific
 Advisory Committee (CASAC)
CASH. *See* Consensus Action on Salt
 and Hypertension (CASH)

Caspari, Ernst, 193–94, 195–98, 203
Castlight Health study, 129
CDC. *See* Centers for Disease Control
 and Prevention (CDC)
Center for Science in the Public
 Interest (CSPI), 78, 93, 95, 99, 100
Centers for Disease Control and
 Prevention (CDC)
 on deaths from hypertension and
 hypertensive renal disease, 128
 efforts to curtail opioid prescriptions, 133, 134
 growth of academic medical
 research and, 164–65
 opioid overdose statistics, 129, 134
 on opioid prescription numbers, 131
 on salt intake and mortality, 110
Central Virginia Seismic Zone
 (CVSZ), 219
CEQ. *See* Council on Environmental
 Quality (CEQ)
Chapman, Carleton, 68
Charles, Donald, 192
Chasis, Herbert, 68–69, 82
chemicals. *See* carcinogens and
 chemicals, regulation of
children, Dahl's anti-salt campaign
 and focus on, 79–81, 86
cholesterol
 current consensus on, 54–56
 debunking of myths surrounding
 dietary, 117–18
 demonization of, 41
 drugs to lower, 168
 Keys's first model and, 44–45
 Keys's second model and, 46–47,
 49–50
 types of, 49–50
Christy, John, 242
Chronicle of Higher Education
 (journal), 33
Church, science and, 9–11
CIA (Central Intelligence Agency),
 MDMA and, 153
Cicero, T. J., 132
cigarette smoking
 heart disease and, 43

Einstein, Albert, 24, 32, 192
EIS. *See* environmental impact
statement (EIS)
Eisenhower, Dwight D., 2, 30–31
Eli Lilly Company, insulin therapy
and, 173
elite, use of peer review by scientific,
1–2
Elliot, Paul, 108
Ely, Daniel, 108
Encyclopédie (Diderot), 20
endangerment finding. *See also*
carbon dioxide
history of, 239–47
Technical Support Document and,
241–242
endorphins, 127
Engels, Friedrich, 13
environmental impact statement
(EIS)
Pebble mine and, 224–25
pressure to switch to cumulative
effects assessments, 227
environmental observational (EO)
study, 275
environmental organizations,
collusion with EPA to prevent
Pebble mine, 224, 225–26, 230
Environmental Protection Agency
(EPA). *See also* carbon dioxide;
fine particulate matter (PM);
Pebble mine
Bristol Bay Watershed (BBWA)
report and, 225, 229–31, 232–36
Cancer Assessment Group, 205–7
CASAC membership, 284–85,
287–89t, 331n16
circumventing NEPA process,
232–33
collusion with environmental
organizations, 224, 225–26, 230
EPA-funded risk production
complex, 280
evidence for causal effect of
exposure to fine PM, 272–74
evidence for 2012 fine PM standard,
264

on fine PM and increase in
mortality, 272
global warming and, 5
hormesis model and, 6–7
how EPA sets air quality standards,
261
LNT model and, 187, 207
long-term cohort studies on fine
PM, 268–72, 271f
model selection bias and, 274–78
Office of Inspector General
investigation on Pebble mine
process, 230–31, 233–36
short-term observational studies on
fine PM, 265–68, 267f, 269f
studies relied on, 284
Trump administration EPA and
Pebble mine site, 236–38
use of Clean Water Act to halt
Pebble mine, 227–29, 231
veto authority and, 228
Environmental Protection Council, 2
environmental risk assessment, 185
EO. *See* environmental observational
(EO) study
EPA. *See* Environmental Protection
Agency (EPA)
EPA Star Grant, 284
epidemiological studies, setting fine
PM standard and, 264
Evans, Robley, 202
evergreening, 133
exercise, hypertension and, 122
experimentation, 32

FACA. *See* Federal Advisory Committee Act (FACA)
"fake science," 224
Fanelli, Daniele, 27, 29, 30
FAO. *See* Food and Agriculture
Organization (FAO)
farmers, investigation of hormetic
relationship between fine PM
and mortality in, 279–80
fat. *See* dietary fat
FDCA (Food, Drug, and Cosmetic
Act of 1938), 155

heroin users, 132
Hesse, Herman, 174
heterogeneity in findings, 273
HHS. *See* Department of Health and
  Human Services (HHS)
hierarchy of evidence, 47–48
High Blood Pressure Education
  Program (NHBPE), 102, 103
Hilleboe, Herman, 41–42
Hippocrates, 9–10
"Historical Hurricane Tracks,
  1933–1998, Virginia and the Caro-
  linas," 213
HLD (high-density lipoprotein),
  49, 51
Hoffmann, Felix, 126
homeostasis, salt intake and, 62,
  63–65, 114–15
Hopke, Philip, 287t, 289t
hormesis model
  relationship between fine PM and
    mortality and, 278–80
  sunlight and, 334n54
  Trump administration and, 6–7
hormetic dose-response model, 186f
Hourdin, Frederic, 6, 249, 250
House Science Subcommittee on
  Oversight, sodium labeling
  hearings, 93–95
Human Genome Project, 21
Hume, David, 14
Hurricane Camille, 214–15, 217f
Hurricane Fran, 216
Hurricane Hazel, 214
hurricanes, *Uranium Mining in
  Virginia* on, 213–17, 215f, 217f
hydrocodone (Vicodin), 127
hydromorphone (Dilaudid), 126–27
Hygienic Laboratory, 164
hypertension. *See also* salt-
  hypertension hypothesis
  age and, 72
  body weight and, 73
  chronic NSAID use and, 128
  DASH diet and, 56
  genetics and, 73
  medications for malignant, 168

potassium and, 119–22
  strategy to reduce, 123
  weight loss and exercise and, 122
Hypertension (journal), 100
hypothetical mine, used in EPA
  assessment of impact of Pebble
  mine, 229–231

ibuprofen, 128
ILDL (large low-density lipoprotein),
  49
imaging techniques, to diagnosis
  heart disease, 168–69
"Immediate Priorities and a Long-
  Range Research Portfolio"
  (NRC), 282
implied corruption, salt debate and
  strategy of, 101
incentive-based bias, government
  policy and, 36
income
  exposure to fine PM and, 333n41
  increase in life expectancy and,
    276–77, 332n33
induction, 32
industrial conspiracy, anti-salt debate
  and claim of, 99–100, 102
industrial copying, costs of, 19–20
industrial knowledge, tacit nature of,
  20
industrial research, lack of free access
  to knowledge, 19–21
Industrial Revolution
  medical innovation and, 162
  scientific laissez faire and, 14, 15
  wealth and private funding of R&D
    and, 19
industrial science, academic science
  and, 12–14
industry bias, 30
infant food, salt in, 79, 80–82, 86
Ingersoll, John, 144
innovation, dissemination of, 20–21.
  See also medical innovation
Insel, Tom, 158
Institute of Medicine (IOM), 103,
  115–16

TREVOR BURRUS is a research fellow in the Cato Institute's Robert A. Levy Center for Constitutional Studies and editor-in-chief of the *Cato Supreme Court Review*. He is also the co-host of *Free Thoughts*, a weekly podcast that covers topics in libertarian theory, history, and philosophy. He is the editor of *A Conspiracy against Obamacare* (Palgrave Macmillan, 2013) and *Deep Commitments: The Past, Present, and Future of Religious Liberty* (Cato, 2017) and holds a BA in philosophy from the University of Colorado at Boulder and a JD from the University of Denver Sturm College of Law.

EDWARD J. CALABRESE is a professor of toxicology at the University of Massachusetts, School of Public Health and Health Sciences, Amherst, and a former adjunct scholar at the Cato Institute. Dr. Calabrese has researched extensively in the area of host factors affecting susceptibility to pollutants and is the author of over 825 papers in scholarly journals, as well as more than 10 books. Dr. Calabrese was awarded the 2009 Marie Curie Prize for his body of work on hormesis. He was the recipient of the International Society for Cell Communication and Signaling-Springer award for 2010. He was awarded an Honorary Doctor of Science Degree from McMaster University in 2013. In 2014 he was awarded the Petr Beckmann Award from Doctors for Disaster Preparedness.

JASON S. JOHNSTON is an adjunct scholar at the Cato Institute and the Henry L. and Grace Doherty Charitable Foundation Professor of Law at the University of Virginia School of Law. He formerly served as Robert G. Fuller Jr. Professor of Law and director of the Program on

Law, Environment and Economy at the University of Pennsylvania Law School. Johnston's scholarship has examined subjects including natural resources law, torts, and contracts (including contractual choice of dispute resolution mechanisms).

NED MAMULA is a former adjunct scholar at the Cato Institute and a senior geoscientist with over 30 years of experience in research and policy development of energy, minerals, and their geopolitics. During his career, Mamula has been employed by leading scientific and intelligence agencies including the U.S. Geological Survey. His focus has been mainly domestic and international energy and mineral research and development projects, including oil and gas, minerals and mining policy, environmental assessments, and the administration of federal leases. He is the author of the recent book *Groundbreaking! America's New Quest for Mineral Independence* (2018). Dr. Mamula received his BA in geology from Slippery Rock University, his MA in geoscience from Penn State University, his PhD in geology and geophysics from Texas A&M University, and an MA in strategic studies and resource policy from Johns Hopkins University's School of Advanced International Studies.

MICHELLE MINTON is a senior fellow at the Competitive Enterprise Institute. Minton specializes in consumer policy, covering regulatory issues that include gambling, tobacco harm reduction, cannabis legalization, alcohol, and nutrition. She has authored numerous studies, including topics like the effectiveness and unintended consequences of sin taxes and history of gambling regulation. Minton holds a BA from Johns Hopkins University and is currently completing her MS in applied nutrition at the University of New England.

JEFFREY A. SINGER is a senior fellow at the Cato Institute and works in the Department of Health Policy Studies. He is principal and founder of Valley Surgical Clinics, Ltd., the largest and oldest group private surgical practice in Arizona. He was integrally involved in the creation and passage of the Arizona Health Care Freedom Act and served as treasurer of the U.S. Health Freedom Coalition, which promotes state constitutional protections of freedom of choice in healthcare decisions. He was a regular contributor to *Arizona Medicine*, the journal of the Arizona Medical Association from 1994 to 2016. He also served on the Advisory Board Council of the Center for Political Thought and Leadership at Arizona State University from 2014 to 2018 and is an adjunct instructor in the Program on Political History and Leadership at ASU. He writes and speaks extensively on regional and national public policy, with a specific focus on the areas of healthcare policy and the harmful effects of drug prohibition. He received his BA from Brooklyn College (CUNY) and his MD from New York Medical College. He is a fellow of the American College of Surgeons.

THOMAS P. STOSSEL is Professor of Medicine Emeritus, Harvard Medical School and the Hematology Division, Brigham Health. He is a member of the National Academy of Sciences, the American Academy of Arts and Sciences, and the National Academy of Medicine, and the founding scientist of BioAegis Therapeutics, a company undertaking clinical development of his discoveries concerning inflammation. Stossel is the author of *Pharmaphobia: How the Conflict of Interest Myth Undermines American Medical Innovation* (Rowman and Littlefield, 2015). With his wife, Kerry Maguire, DDS, MSPH, Stossel founded Options for Children in Zambia that provides healthcare services in that Sub-Saharan African country.

TERENCE KEALEY is a visiting senior fellow at the Cato Institute and a professor of clinical biochemistry at the University of Buckingham in the United Kingdom, where he served as vice chancellor until 2014. He is the author of *The Economic Laws of Scientific Research* and *Breakfast Is a Dangerous Meal: Why You Should Ditch Your Morning Meal for Health and Wellbeing*. Professor Kealey trained initially in medicine at Bart's Hospital Medical School, London. He studied for his doctorate at Oxford University, where he worked first as a Medical Research Council Training Fellow and then as a Wellcome Senior Research Fellow in Clinical Science.

PATRICK J. MICHAELS is a senior fellow with the Competitive Enterprise Institute and former director of the Center for the Study of Science at the Cato Institute. Michaels is a past president of the American Association of State Climatologists and was program chair for the Committee on Applied Climatology of the American Meteorological Society. He was a research professor of Environmental Sciences at University of Virginia for 30 years. He is the author or editor of six books on climate and its impact, and he was an author of the climate "paper of the year" awarded by the Association of American Geographers in 2004. Michaels holds AB and SM degrees in biological sciences and plant ecology from the University of Chicago, and he received a PhD in ecological climatology from the University of Wisconsin at Madison in 1979.

Founded in 1977, the Cato Institute is a public policy research foundation dedicated to broadening the parameters of policy debate to allow
consideration of more options that are consistent with the principles of
limited government, individual liberty, and peace. To that end, the Institute strives to achieve greater involvement of the intelligent, concerned
lay public in questions of policy and the proper role of government.

The Institute is named for *Cato's Letters*, libertarian pamphlets that
were widely read in the American Colonies in the early 18th century
and played a major role in laying the philosophical foundation for the
American Revolution.

Despite the achievement of the nation's Founders, today virtually
no aspect of life is free from government encroachment. A pervasive
intolerance for individual rights is shown by government's arbitrary
intrusions into private economic transactions and its disregard for civil
liberties. And while freedom around the globe has notably increased
in the past several decades, many countries have moved in the opposite
direction, and most governments still do not respect or safeguard the
wide range of civil and economic liberties.

To address those issues, the Cato Institute undertakes an extensive
publications program on the complete spectrum of policy issues. Books,
monographs, and shorter studies are commissioned to examine the federal budget, Social Security, regulation, military spending, international
trade, and myriad other issues. Major policy conferences are held
throughout the year, from which papers are published thrice yearly in
the *Cato Journal*. The Institute also publishes the quarterly magazine
Regulation.

In order to maintain its independence, the Cato Institute accepts no
government funding. Contributions are received from foundations,
corporations, and individuals, and other revenue is generated from
the sale of publications. The Institute is a nonprofit, tax-exempt, educa-
tional foundation under Section 501(c)3 of the Internal Revenue Code.

CATO INSTITUTE
1000 Massachusetts Ave., NW
Washington, DC 20001
www.cato.org